EXERCISE ANSWERS

math ESSENTIALS

Finding & Filling the Gaps

HERON BOOKS
K-12 CURRICULUM

Published by
Heron Books, Inc.
20950 SW Rock Creek Road
Sheridan, OR 97378

heronbooks.com

———————

Special thanks to all the teachers and students who
provided feedback instrumental to this edition.

———————

Third Edition © 2014, 2019, Heron Books
All Rights Reserved

Spiralbound ISBN: 978-0-89739-129-0
Paperback ISBN: 978-0-89-739200-6

Any unauthorized copying, translation, duplication or distribution, in whole or in part,
by any means, including electronic copying, storage or transmission,
is a violation of applicable laws.

Printed in the USA

25 October 2019

Contents

1. Answers for Place Value 1
2. Answers for Division 7
3. Answers for Word Problems: Whole Numbers 13
4. Answers for Fractions: Adding & Subtracting 19
5. Answers for Fractions: Multiplying & Dividing 29
6. Answers for Word Problems: Fractions 35
7. Answers for Decimals 41
8. Answers for Metric Measurement 57
9. Answers for Customary Measurement 65
10. Answers for Positive and Negative Numbers 71
11. Answers for Simple Algebra 89
12. Answers for Ratio, Proportion and Percent 95
13. Answers for Simple Geometry 105
14. Answers for Supplemental Exercise Sheets 115

ANSWERS FOR
Place Value

Exercise 1

1. Thousands, 6,000
2. Hundreds, 0 hundreds
3. Tens, 80
4. Ten thousands, 20,000
5. Ones, 2
6. Thousands, 0 thousands
7. Ten thousands, 50,000

Additional Exercises

1. Hundreds, 700
2. Ones, 3
3. Tens, 40
4. Thousands, 7,000
5. Hundreds, 500
6. Tens, 10
7. Ones, 9
8. Thousands, 9,000
9. Hundreds, 400
10. Tens, 60
11. Ones, 0 ones
12. Tens, 0 tens
13. Hundreds, 500
14. Ones, 7
15. Ten thousands, 30,000
16. Thousands, 9,000
17. Ten thousands, 50,000

Exercise 2

1. 674,205
2. 109,814
3. 4,862,579
4. 85,761,551
5. 991,030,115
6. 12,200,347
7. 813,000,259
8. 3,637,458,934
9. 47,125,796,809
10. 854,238,721,400
11. 999,600,846,030
12. 50,200,000
13. Four hundred nine thousand seven hundred sixty-five
14. Nine hundred fifty thousand four hundred forty-four
15. One million five hundred eighty-nine thousand four hundred
16. Fifty-two million two hundred sixty-nine thousand one hundred ninety-six
17. Five hundred two million two hundred eighty-one thousand seven
18. Four hundred sixty-five million four thousand one hundred twenty
19. Three million fifty thousand
20. Seven billion nine hundred ninety-nine million four hundred forty-seven thousand two hundred
21. Sixty-one billion four hundred ninety-seven million five hundred thirteen thousand seventy
22. Seven hundred seventy-seven billion four hundred nine million three hundred fifty thousand eleven
23. Eight hundred two billion five hundred forty-three million five hundred thousand

Additional Exercises

1. 523,462
2. 411,391
3. 1,726,985
4. 46,534,111
5. 138,721,873
6. 513,408,067

7. 2,000,674

8. 402,057,020

9. 514,001,000

10. 1,298,482,655

11. 75,613,908,747

12. 542,629,357,940

13. 610,900,743,090

14. 505,800,001,000

15. 750,000,000,000

16. Seven hundred sixty-nine thousand five hundred seventy-two

17. Two hundred thirty-five thousand five hundred eighty-five

18. Three hundred thirty thousand

19. Two million five hundred nineteen thousand four hundred sixty-eight

20. Six million two hundred fifty-four thousand

21. Thirty-six million three hundred ninety-four thousand five hundred eighty-two

22. Six hundred forty-six million nine hundred eighty-three thousand two hundred twenty-nine

23. Eight hundred seven million sixty thousand four hundred fifty-seven

24. Two hundred fifty million four hundred thousand one

25. Ten million

26. Four hundred three million

27. Seven billion five hundred million

28. Seventeen billion four hundred ninety-six million three hundred thousand

29. Two hundred fifty-six billion six hundred seventy-nine million five hundred eleven thousand

30. Seven hundred four billion five hundred million seven hundred fifty thousand three hundred eighty-two

31. Five hundred billion three hundred thirteen thousand

32. Fifty million one hundred seventy thousand

33. Thirteen million four hundred thousand

Exercise 3

1. 90
2. 4,900
3. 707,200
4. 2,000
5. 10,000
6. 20,000
7. 40,000
8. 5,500,000
9. 600,000
10. 14,000,000
11. 10,000,000
12. 52,000,000
13. Already rounded to nearest ten thousand

Additional Exercises

1. 20
2. 80
3. 200
4. 7,540
5. 800
6. 3,000
7. 0
8. 4,000
9. 8,000
10. 6,000
11. Already rounded to nearest thousand
12. 42,000
13. 77,000
14. 699,900
15. 700,000
16. 530,000
17. 527,000
18. 527,500
19. 16,000,000
20. 14,000,000
21. 14,050,000
22. 14,046,000
23. 36,100,000
24. 36,000,000
25. 40,000,000
26. 10,000,000

Final Unit Exercise

1. Hundred thousands
2. Ten billions
3. 0
4. 1
5. 2,084,912
6. 25,771,862,319
7. Two million eight thousand five hundred sixty-six
8. Four hundred eighty-six billion three hundred million nine hundred fifty-one thousand two hundred twenty-three
9. 754,000
10. 3,000
11. 57,190,000

ANSWERS FOR Division

Review Exercise

1. 9 ten thousands
   4 thousands
   0 hundreds
   1 tens
   2 ones

2. 80,739
   Unit 1 Place Value, chapter *The Place Value System*

3. 19,645,300
   Unit 1 Place Value, chapter *Place Values for Larger Numbers*

4. 350

5. 300

6. 79,000
   Unit 1 Place Value, chapter *Rounding Numbers*

7. 720,001

8. 499,203

9. 57,686,688
   Instructor to determine the appropriate assignment.

Exercise 1

1. 2R1
2. 2R2
3. 1R3
4. 2
5. 2R1
6. 2R5
7. 3
8. 1
9. 12
10. 6R2
11. 1
12. 4R2
13. 2R3
14. 11

Additional Exercises

1. 1R2
2. 1R1
3. 1R2
4. 1R4
5. 2
6. 2R1
7. 3R3
8. 4
9. 4R3
10. 5
11. 19
12. 1
13. 9
14. 8R1
15. 6R1
16. 4R1
17. 1
18. 3

Exercise 2

1. 5R1
2. 6R3
3. 163R4
4. 52R2
5. 1,595
6. 657R3
7. 18,854

Additional Exercises

1. 7R1
2. 3R5
3. 11
4. 11R3
5. 7R6
6. 183
7. 237R1
8. 73R3
9. 37R1
10. 1,364R3
11. 342R3
12. 12,693R1
13. 6,736R4

Exercise 3

1. 43R10
2. 417R11
3. 34R6
4. 17

Additional Exercises

1. 26R11
2. 14
3. 222R2
4. 42R1
5. 21R18

Exercise 4

1. 6
2. 7R2
3. 8
4. 6R3
5. 8R2
6. 7R22
7. 15
8. 45R8
9. 941R22
10. 1,287R47

Additional Exercises

1. 8
2. 7
3. 7R56
4. 6R17
5. 9R2
6. 7R29
7. 11R1
8. 42
9. 117R42
10. 941R17
11. 1,392R21
12. 681

Exercise 5

1. 307
2. 60R22
3. 240R7
4. 206R12
5. 706R10
6. 63R15
7. 4,007

Additional Exercises

1. 408
2. 90R21
3. 405
4. 23R2
5. 207
6. 304R2
7. 130R9
8. 307R3
9. 72R24
10. 3,009

Exercise 6

1. 123
2. 394
3. 806R19
4. 1495
5. 607R1
6. 732R101
7. 540R3,017

Additional Exercises

1. 241
2. 725R29
3. 482R5
4. 508R200
5. 289
6. 1158R27
7. 601R786
8. 724
9. 497R71
10. 446R9,432

Exercise 7

1. Undefined
2. 0
3. Undefined
4. 0
5. Undefined
6. Undefined
7. 0
8. 1
9. Undefined
10. 1
11. 0

Additional Exercises

1. Undefined
2. 0
3. Undefined
4. 0
5. Undefined
6. Undefined
7. 0
8. 1
9. Undefined
10. 1
11. 0

Exercise 8

1. 12 points per game
2. 3 miles each day average
3. 7 miles per hour
4. 17,160 people
5. 15 pages per hour
6. 56R3, rounded up to 57 years
7. 4 gallons per minute
8. $42,078

Additional Exercises

1. 24 pounds per dog
2. 262 students in each school
3. 174 pounds
4. 190 miles per day
5. $9 per hour
6. 25 words per minute
7. 17,886 average population

Final Unit Exercises

1. 627R2
2. 486R21
3. 807R46
4. 3,019R90
5. 483R3,473
6. 0
7. 2,006
8. Undefined
9. 49 miles per hour
10. 99R1, rounded to the nearest point is 99

ANSWERS FOR Word Problems: Whole Numbers

Exercise 1

1. 193 minutes
2. 254 miles
3. $180
4. 185 people
5. $7
6. 9 pounds
7. 111 minutes
8. 25,786 tickets
9. 731 seats
10. 18 games

Additional Exercises

1. $891
2. 522 miles
3. 79,240 spectators
4. $3,449
5. $28
6. 225 minutes
7. 12,769 more people

Exercise 2

1. $228
2. 450 players
3. 114 miles
4. $126,900
5. 22,425 passengers
6. $18,900
7. $337,500

Additional Exercises

1. 416 seats
2. $360
3. 420 gallons
4. 1,696 players
5. 35,140 miles
6. $74,970

Exercise 3

1. $217
2. $884,354
3. $216
4. 427 people
5. $798
6. 46,334 miles
7. 6,110 miles
8. $143
9. 646 boxes
10. 804 hours
11. Yes; $2,250 more

Additional Exercises

1. $63
2. $979
3. $1,041,593
4. 448 miles
5. 3,479 passengers
6. $176
7. 530 miles
8. 827 miles
9. $3,160
10. 19 people

Exercise 4

1. $87
2. 222 boxes
3. $88,825
4. 31 days
5. $15,482 R $3 ($15,482.25)
6. $208 R $4 ($208.80)
7. 54 miles per hour
8. 407 desks, with a remainder of $389
9. $980 R $4, ($980.80)
10. 74 years
11. 395 houses
12. 30 remainder 7, rounded down to 30 pages per hour
13. 58 boxes

Additional Exercises

1. 763 days
2. $520,500
3. 7,680 people
4. $6,507 R $273 ($6,507.97)
5. 13 trips R 50 people, rounded up to 14 trips
6. 25 rows
7. 4,507R15, rounded down to 4,507 people
8. 40,556R2, rounded up to 40,557 average seats
9. 35 miles R 150, rounded down to 35 miles
10. $75
11. 24 people R 100 pounds, rounded down to 24 people

Exercise 5

1. 5,460 hours
2. 75 pounds
3. 10,080 minutes
4. 264 hours total; 25 miles per hour R 200, rounded up to 26
5. 1,950 minutes
6. 6 miles R 8, rounded down to 6 miles
7. $7,436,625
8. 31 hours
9. 6,480 people
10. 13 points

Additional Exercises

1. $242,000
2. 253 pounds
3. $740,025 collected; $759 per tree
4. 39 rows
5. 56,904 miles
6. 9,282R2, rounded up to 9,283 times
7. 400 days
8. $111,076
9. 1,321R2, rounded down to 1,321 miles
10. 609R16, rounded down to 609 pounds

Exercise 6

1. 352 hours
2. 2,380 gallons per day; 140 gallons per hour
3. 310 miles
4. $19,110
5. $9,110
6. $135
7. $30
8. 83 average score
9. $256
10. $727,122
11. 7 days
12. 7R17, rounded down to 7 days
13. $391
14. 74 miles per hour
15. 84 points

Additional Exercises

1. 965 customers
2. $29,802,750
3. Golfer A, his average was about 321 yards
4. 24 questions
5. 880 miles
6. 16 hours
7. $35,000
8. 36 small pizzas; $432
9. Yes; $24
10. 32,800 more; 4,100 views
11. 16,000 points
12. 6 points
13. Finley, Finley 231, Stan 229

ANSWERS FOR
Fractions: Adding & Subtracting

Review Exercise

1. 24,136

Instructor to determine the appropriate assignment

2. $641,330

Instructor to determine the appropriate assignment

3. 187

Unit 2 Division, chapter *Averages*

4. 7,008

Unit 2 Division, chapter *Zeros in the Quotient*

Worksheet Exercise

1.

2.

3.

4.

5.

6.

7.

8.

9.

10.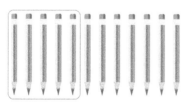

11.

12.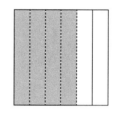

13. $\frac{3}{7}$

14. $\frac{4}{7}$

15. $\frac{7}{9}$

16. $\frac{2}{9}$

17. $\frac{3}{16}$

18. $\frac{13}{16}$

Exercise 1

1. $\frac{3}{9}$
2. $\frac{6}{21}$
3. $\frac{4}{24}$
4. $\frac{9}{21}$
5. $\frac{6}{8}$
6. $\frac{21}{24}$
7. $\frac{4}{5}$
8. $\frac{1}{3}$
9. $\frac{4}{5}$
10. $\frac{3}{8}$
11. $\frac{2}{5}$
12. $\frac{9}{10}$
13. $\frac{2}{16}$ or equivalent
14. $\frac{10}{32}$ or equivalent
15. $\frac{6}{16}$ or equivalent
16. $\frac{14}{30}$ or equivalent
17. $\frac{4}{6}$ or equivalent
18. $\frac{2}{4}$ or equivalent
19. $\frac{2}{9}$
20. $\frac{1}{3}$
21. $\frac{5}{9}$
22. $\frac{3}{5}$
23. $\frac{1}{2}$
24. $\frac{2}{11}$
25. $\frac{3}{4}$
26. $\frac{18}{95}$
27. $\frac{3}{8}$
28. $\frac{1}{7}$
29. $\frac{2}{3}$
30. $\frac{5}{7}$

Additional Exercises

1. $\frac{3}{12}$
2. $\frac{4}{14}$
3. $\frac{28}{32}$
4. $\frac{4}{10}$
5. $\frac{6}{21}$
6. $\frac{3}{9}$
7. $\frac{6}{8}$
8. $\frac{14}{16}$
9. $\frac{10}{32}$
10. $\frac{15}{50}$
11. $\frac{3}{4}$
12. $\frac{5}{8}$
13. $\frac{1}{2}$
14. $\frac{3}{5}$
15. $\frac{7}{8}$
16. $\frac{1}{4}$
17. $\frac{2}{3}$
18. $\frac{1}{2}$
19. $\frac{9}{10}$
20. $\frac{4}{5}$
21. $\frac{2}{10}$ or equivalent
22. $\frac{14}{24}$ or equivalent
23. $\frac{6}{14}$ or equivalent
24. $\frac{6}{32}$ or equivalent
25. $\frac{10}{18}$ or equivalent
26. $\frac{5}{11}$
27. $\frac{3}{7}$
28. $\frac{7}{20}$
29. $\frac{2}{5}$
30. $\frac{3}{5}$
31. $\frac{3}{4}$
32. $\frac{1}{4}$
33. $\frac{2}{3}$
34. $\frac{5}{6}$
35. $\frac{9}{16}$

Exercise 2

1. 2, 4, 6, 8, 10, 12, 14
2. 4, 8, 12, 16, 20, 24
3. 6, 12, 18, 24, 30, 36
4. 8, 16, 24, 32, 40, 48
5. 7, 14, 21, 28 and 35
6. 9, 18, 27, 36 and 45
7. 10
8. 9
9. 21
10. 56
11. 143
12. 15, $\frac{5}{15}$, $\frac{6}{15}$
13. 9, $\frac{3}{9}$, $\frac{2}{9}$
14. 16, $\frac{3}{16}$, $\frac{4}{16}$
15. 12, $\frac{9}{12}$, $\frac{10}{12}$
16. 60, $\frac{8}{60}$, $\frac{9}{60}$
17. 60, $\frac{25}{60}$, $\frac{16}{60}$
18. 77, $\frac{22}{77}$, $\frac{14}{77}$
19. $\frac{5}{10}$ > $\frac{4}{10}$ so $\frac{1}{2}$ > $\frac{2}{5}$
20. $\frac{8}{12}$ < $\frac{9}{12}$ so $\frac{2}{3}$ < $\frac{3}{4}$
21. $\frac{24}{40}$ < $\frac{25}{40}$ so $\frac{3}{5}$ < $\frac{5}{8}$
22. $\frac{15}{18}$ > $\frac{14}{18}$ so $\frac{5}{6}$ > $\frac{7}{9}$
23. $\frac{4}{24}$ > $\frac{3}{24}$ so $\frac{1}{6}$ > $\frac{1}{8}$
24. $\frac{39}{45}$ > $\frac{35}{45}$ so $\frac{13}{15}$ > $\frac{7}{9}$
25. $\frac{209}{285}$ > $\frac{180}{285}$ so $\frac{11}{15}$ > $\frac{12}{19}$
26. $\frac{12}{21}$ > $\frac{11}{21}$ so $\frac{4}{7}$ > $\frac{11}{21}$

Additional Exercises

1. 3, 6, 9, 12, 15, 18, 21
2. 5, 10, 15, 20, 25, 30
3. 7, 14, 21, 28, 35, 42
4. 9, 18, 27, 36, 45, 54
5. 8, 16, 24, 32, 40, 48, 56
6. 11, 22, 33, 44, 55, 66, 77, 110, 121
7. 15
8. 12
9. 8
10. 30
11. 36
12. 60
13. 72

14. 28

15. 12, $\frac{3}{12}$, $\frac{8}{12}$

16. 20, $\frac{5}{20}$, $\frac{4}{20}$

17. 12, $\frac{9}{12}$, $\frac{2}{12}$

18. 9, $\frac{6}{9}$, $\frac{2}{9}$

19. 24, $\frac{8}{24}$, $\frac{9}{24}$

20. 63, $\frac{27}{63}$, $\frac{7}{63}$

21. 48, $\frac{32}{48}$, $\frac{9}{48}$

22. 100, $\frac{6}{100}$, $\frac{35}{100}$

23. $\frac{9}{24} > \frac{4}{24}$ so $\frac{3}{8} > \frac{1}{6}$

24. $\frac{3}{6} > \frac{2}{6}$ so $\frac{1}{2} > \frac{1}{3}$

25. $\frac{5}{10} < \frac{6}{10}$ so $\frac{1}{2} < \frac{3}{5}$

26. $\frac{15}{20} < \frac{16}{20}$ so $\frac{3}{4} < \frac{4}{5}$

27. $\frac{6}{30} > \frac{5}{30}$ so $\frac{1}{5} > \frac{1}{6}$

28. $\frac{25}{80} < \frac{36}{80}$ so $\frac{5}{16} < \frac{9}{20}$

29. $\frac{33}{77} < \frac{35}{77}$ so $\frac{3}{7} < \frac{5}{11}$

30. $\frac{27}{144} < \frac{40}{144}$ so $\frac{3}{16} < \frac{5}{18}$

Exercise 3

1. $\frac{2}{5}$

2. $\frac{1}{2}$

3. $\frac{1}{2}$

4. $\frac{11}{12}$

5. $\frac{17}{24}$

6. $\frac{5}{7}$

7. $\frac{1}{4}$

8. $\frac{1}{4}$

9. $\frac{1}{6}$

10. $\frac{4}{9}$

11. $\frac{1}{2}$

12. $\frac{21}{40}$

13. $\frac{7}{36}$

14. $\frac{69}{80}$

15. $\frac{13}{56}$

16. $\frac{131}{260}$

17. $\frac{5}{14}$ mile

18. $\frac{9}{16}$ cup

19. $\frac{31}{48}$ ounce

20. $\frac{37}{48}$ pound

Additional Exercises

1. $\frac{2}{3}$
2. $\frac{2}{3}$
3. $\frac{3}{4}$
4. $\frac{5}{6}$
5. $\frac{7}{18}$
6. $\frac{3}{5}$
7. $\frac{5}{8}$
8. $\frac{8}{9}$
9. $\frac{20}{21}$
10. $\frac{1}{3}$
11. $\frac{1}{2}$
12. $\frac{1}{8}$
13. $\frac{1}{4}$
14. $\frac{1}{2}$
15. $\frac{1}{12}$
16. $\frac{11}{24}$
17. $\frac{1}{24}$
18. $\frac{32}{63}$
19. $\frac{17}{60}$
20. $\frac{20}{91}$
21. $\frac{1}{6}$
22. $\frac{56}{221}$
23. $\frac{1}{36}$
24. $\frac{1}{12}$ cup
25. $\frac{11}{12}$ cup
26. $\frac{5}{24}$ mile
27. $\frac{23}{24}$ pound

Exercise 4

1. $1\frac{2}{5}$
2. $1\frac{1}{2}$
3. $4\frac{17}{24}$
4. $6\frac{15}{16}$
5. $10\frac{5}{16}$
6. $7\frac{20}{77}$
7. $3\frac{2}{3}$
8. $8\frac{5}{8}$
9. $9\frac{9}{20}$
10. $9\frac{7}{16}$
11. $2\frac{11}{15}$
12. $5\frac{7}{12}$
13. $2\frac{2}{3}$
14. $6\frac{5}{12}$
15. $18\frac{17}{24}$
16. $61\frac{47}{110}$
17. 26
18. $17\frac{9}{16}$
19. $78\frac{1}{6}$
20. $11\frac{14}{17}$
21. $549\frac{3}{10}$
22. $4,497\frac{35}{36}$
23. $1,980\frac{7}{16}$ gallons
24. $12\frac{7}{12}$ pounds

Additional Exercises

1. $1\frac{1}{3}$
2. $1\frac{1}{3}$
3. 1
4. $4\frac{1}{2}$
5. $5\frac{7}{12}$
6. $13\frac{27}{40}$
7. $20\frac{5}{8}$
8. $5\frac{23}{42}$
9. $16\frac{8}{15}$
10. $929\frac{5}{12}$
11. $1\frac{2}{3}$

12. $6\frac{5}{12}$

13. 7

14. $1\frac{3}{16}$

15. $6\frac{13}{24}$

16. $1\frac{2}{3}$

17. $2\frac{5}{8}$

18. $1\frac{11}{24}$

19. $1\frac{1}{4}$

20. $5\frac{1}{8}$

21. $5\frac{5}{12}$

22. $6\frac{7}{10}$

23. $32\frac{17}{24}$

24. $772\frac{20}{33}$

25. $1\frac{37}{60}$

26. $67\frac{8}{11}$

27. $489\frac{17}{48}$

28. $670\frac{43}{80}$

29. $20\frac{5}{8}$ pounds

30. $4\frac{17}{24}$ pounds

31. $2\frac{4}{5}$ miles

32. $25\frac{7}{12}$ gallons

Final Unit Exercises

1. $\frac{2}{7}$

2. $\frac{7}{18} > \frac{5}{16}$

3. $\frac{11}{144}$

4. $300\frac{6}{35}$

5. $121\frac{17}{36}$

6. $24\frac{11}{16}$ pounds

7. $7\frac{5}{24}$ pounds

ANSWERS FOR Fractions: Multiplying & Dividing

Review Exercise

1. 3,008

   Unit 2 Division, chapter Zeros in the Quotient

2. Answers will vary, for example $\frac{6}{14}, \frac{9}{21}$

   Unit 4 Fractions Adding & Subtracting, chapter Equivalent Fractions

3. $101\frac{13}{20}$

   Unit 4 Fractions Adding & Subtracting, chapter Adding and Subtracting Mixed Numbers

4. $29\frac{7}{12}$

   Unit 4 Fractions Adding & Subtracting, chapter Adding and Subtracting Mixed Numbers

5. 8 boats, because 7 boats carries only 1855, not 2000.

   Unit 2 Division, chapter Three-Digit Divisors, section Zero and Dividing

6. 756,000 paper clips

   Instructor to determine the appropriate assignment

Exercise 1

1. 8
2. $12\frac{1}{2}$
3. $\frac{1}{4}$
4. $\frac{7}{25}$
5. $\frac{9}{35}$
6. $\frac{1}{6}$
7. 18
8. 16
9. $12\frac{1}{2}$
10. $\frac{8}{15}$
11. $\frac{3}{10}$
12. $\frac{5}{32}$
13. 19 pounds
14. 7 shots
15. 87 students
16. $\frac{1}{4}$ mile
17. $\frac{7}{12}$ pound

Additional Exercises

1. $8\frac{1}{2}$
2. 1
3. $3\frac{1}{3}$
4. 7
5. $\frac{3}{16}$
6. $\frac{1}{20}$
7. $3\frac{3}{4}$
8. $\frac{7}{10}$
9. $\frac{1}{4}$
10. $\frac{3}{4}$
11. $\frac{2}{25}$
12. 18
13. $\frac{1}{9}$
14. $\frac{3}{10}$
15. $\frac{5}{32}$
16. 8 students
17. $\frac{7}{16}$ cup
18. $2\frac{2}{3}$ miles
19. 28 books
20. $\frac{9}{20}$ mile

Exercise 2

1. $\frac{13}{4}$
2. $\frac{23}{8}$
3. $\frac{23}{6}$
4. $\frac{286}{7}$
5. $\frac{38}{9}$
6. $\frac{98}{5}$
7. $18\frac{3}{4}$
8. $1\frac{11}{16}$
9. $8\frac{14}{27}$
10. $32\frac{2}{3}$
11. $66\frac{5}{8}$
12. $166\frac{19}{20}$
13. $26\frac{1}{4}$ inches
14. $1\frac{1}{4}$ cups
15. $5\frac{5}{6}$ cups

Additional Exercises

1. $\frac{11}{2}$
2. $\frac{53}{8}$
3. $\frac{17}{3}$
4. $\frac{23}{10}$
5. $\frac{38}{3}$
6. $\frac{121}{5}$
7. $\frac{158}{9}$
8. $\frac{151}{8}$
9. $\frac{197}{6}$
10. $\frac{217}{16}$
11. $6\frac{3}{5}$
12. 5
13. $3\frac{23}{32}$
14. $25\frac{13}{20}$
15. $12\frac{1}{2}$
16. $69\frac{25}{48}$
17. $3\frac{1}{2}$ hours
18. $30\frac{5}{8}$ hours
19. $10\frac{7}{15}$ feet
20. $8\frac{3}{4}$ cups

Exercise 3

1. a)

   b) 6

   c) 6

   d) 6

2. a)

   b) 6

   c) 6

   d) 6

3. 2

4. $\frac{9}{2}$

5. $\frac{1}{16}$

6. $\frac{4}{17}$

7. $\frac{1}{25}$

8. 2

9. $\frac{16}{5}$

10. $\frac{1}{8}$

11. $\frac{5}{7}$

12. $\frac{1}{5}$

13. $\frac{5}{2}$

14. $\frac{1}{50}$

15. 24

16. 20

17. 45

18. 234

19. 100

20. $23\frac{1}{3}$

Additional Exercises

1. a)

   b) 4

   c) 4

   d) 4

2. a)

   b) 3

    c) 3

    d) 3

3. 7
4. $\frac{15}{2}$
5. $\frac{1}{13}$
6. $\frac{3}{19}$
7. $\frac{1}{6}$
8. 4
9. $\frac{5}{2}$
10. $\frac{1}{3}$
11. $\frac{1}{10}$
12. $\frac{4}{3}$
13. $\frac{3}{5}$
14. $\frac{1}{11}$
15. $\frac{1}{100}$
16. 15
17. 12
18. 12
19. 60
20. 200
21. 208

Exercise 4

1. $3\frac{1}{2}$
2. $2\frac{12}{29}$
3. $5\frac{1}{9}$
4. $\frac{1}{11}$
5. $1\frac{1}{5}$
6. 5
7. $4\frac{2}{3}$
8. $4\frac{5}{12}$
9. $2\frac{2}{5}$ miles per hour average
10. 12 servings and no cereal left over
11. $2\frac{11}{16}$ miles per hour average
12. $8\frac{27}{32}$ ounces, average weight
13. 11 people, $\frac{3}{4}$ ounce of chocolate left over
14. 75 people, no granola left over
15. 15 pieces, $1\frac{1}{4}$ inches of sandwich left over

Additional Exercises

1. $3\frac{8}{9}$
2. $2\frac{6}{17}$
3. $5\frac{7}{32}$
4. $\frac{1}{11}$
5. $2\frac{1}{2}$
6. $4\frac{2}{3}$
7. $2\frac{2}{39}$
8. $2\frac{3}{128}$
9. 7
10. $3\frac{23}{27}$
11. $13\frac{1}{3}$ pages per hour
12. 26 people, $\frac{1}{4}$ pizza left over
13. 7 cakes, $3\frac{1}{16}$ pounds left over
14. 64 cups, none left over
15. $15\frac{7}{16}$ pounds
16. 6 swords, 3 feet left over
17. 6 glasses, $2\frac{3}{4}$ ounces left over
18. $27\frac{1}{2}$ miles per hour
19. $13\frac{1}{11}$ miles per hour

Final Unit Exercise

1. 56
2. $\frac{5}{32}$
3. $\frac{7}{72}$
4. $11\frac{1}{7}$
5. 152
6. $\frac{171}{10}$, reciprocal is $\frac{10}{171}$
7. $\frac{1}{1,000}$
8. 2
9. 20
10. $110\frac{14}{51}$
11. $4\frac{9}{16}$
12. 24 people, $2\frac{3}{4}$ ounces left over
13. 32 people, no spaghetti left over

ANSWERS FOR Word Problems: Fractions

Exercise 1

1. $2\frac{5}{8}$ cups
2. $3\frac{9}{20}$ miles
3. $33\frac{5}{8}$ pounds
4. $\frac{5}{8}$ of a pound
5. $\frac{13}{30}$ feet
6. $\frac{3}{5}$ of the recipe
7. yes
8. $4\frac{239}{240}$ gallons

Additional Exercises

1. $1\frac{1}{5}$ hour
2. $14\frac{7}{8}$ inches
3. $\frac{5}{24}$ gallon
4. $\frac{5}{16}$
5. $1\frac{1}{8}$ feet
6. $10\frac{5}{12}$ pounds
7. $\frac{5}{8}$ completed
8. $21\frac{11}{16}$ gallons
9. $7\frac{1}{2}$ ounces
10. $68\frac{11}{48}$ feet
11. $\frac{1}{3}$ of the recipe
12. $6\frac{9}{16}$ pound

Exercise 2

1. $47\frac{1}{4}$ ounces
2. $18\frac{3}{4}$ pounds
3. $\frac{11}{15}$ mile
4. $2\frac{5}{8}$ cups
5. $8\frac{13}{16}$ miles
6. no, it is $\frac{1}{15}$ bigger
7. $7\frac{7}{24}$ cups
8. 3,300 feet in $\frac{5}{8}$ mile
9. $\frac{5}{24}$ gallon
10. $\frac{7}{12}$ gallon used; $\frac{7}{24}$ gallon left
11. $4\frac{3}{5}$ gallons
12. 42 friends sent to; 14 friends not yet sent to
13. $21\frac{1}{2}$ gallons to sell; 14 gallons to make

Additional Exercises

1. $21
2. 40 cookies
3. 15 questions
4. $15\frac{1}{8}$ meters
5. 2 eggs
6. 9 eggs used; 3 eggs left
7. $38\frac{1}{2}$ feet
8. 24 shots
9. 13,200 feet
10. 2,640 feet
11. $11\frac{1}{2}$ feet
12. $4\frac{17}{20}$ pounds
13. $2\frac{1}{10}$ pounds
14. $\frac{7}{16}$ ounce
15. $5\frac{1}{3}$ miles
16. walked 4 miles; 2 miles left to walk
17. walked $\frac{15}{64}$ miles; $\frac{45}{64}$ mile further
18. $\frac{1}{16}$ mile
19. $9\frac{5}{6}$ pounds used; $4\frac{11}{12}$ pounds left

Exercise 3

1. $4\frac{1}{2}$ miles per hour
2. $4\frac{13}{17}$ miles per hour
3. 95 minutes
4. $24\frac{8}{9}$ hours
5. $\frac{1}{3}$ of the recipe; $3\frac{7}{12}$ quarts
6. 44 trick or treaters
7. 32 ounces
8. $6\frac{13}{30}$ miles
9. $112\frac{1}{8}$ pages
10. $37\frac{31}{32}$ servings
11. $46\frac{7}{8}$ ounces
12. $15\frac{5}{21}$ ounces
13. $3\frac{13}{35}$ hours
14. $54\frac{22}{47}$ cups
15. $85\frac{1}{3}$ packets

Additional Exercises

1. 7 hours
2. 20 hours
3. 35 quarts
4. 12 servings
5. 12 guests
6. 72 ounces
7. 6 ounces
8. $\frac{2}{3}$; 2 cups
9. 8 servings
10. 15 pages per hour
11. $13\frac{1}{2}$ pages per hour
12. 162 ounces
13. 20 servings
14. $\frac{3}{8}$ cup
15. $7\frac{1}{2}$ miles
16. $7\frac{53}{60}$ miles
17. 5 hours
18. 4 or $\frac{4}{1}$; 8 quarts
19. 8 months
20. $6\frac{4}{43}$ hours
21. $2\frac{1}{2}$; 5 pounds

Exercise 4

1. 33 people; 1 cup

2. 6 pizzas

3. $4\frac{21}{32}$ pounds to peel; $2\frac{15}{32}$ pounds left to peel after her friend helps.

4. 40 people can be served; $\frac{2}{3}$ cup left over.

5. $\frac{5}{6}$ teaspoon yeast

   2 tablespoons water

   $1\frac{1}{6}$ cups flour

   $\frac{1}{2}$ cup milk, rounded to $\frac{4}{8} = \frac{1}{2}$

   $1\frac{1}{3}$ tablespoons vegetable oil

   $\frac{7}{12}$ cup butter, rounded to $\frac{8}{12} = \frac{2}{3}$ cup

6. $3\frac{1}{8}$ teaspoons yeast

   $7\frac{1}{2}$ tablespoons water

   $4\frac{3}{8}$ cup flour

   5 tablespoons vegetable oil

   $2\frac{3}{16}$ cups butter. This could be rounded to $2\frac{4}{16} = 2\frac{1}{4}$ cups.

   $2\frac{1}{2}$ eggs. This could be rounded to 3 eggs.

7. $6\frac{1}{4}$ miles

8. $35\frac{5}{7}$ times, or 35 times

Additional Exercises

1. $\frac{3}{4}$ of the games won; $\frac{1}{4}$ of the games lost.
2. 18 quarts
3. 3 servings
4. $\frac{9}{16}$ cup brown sugar. Rounded off to $\frac{8}{16} = \frac{1}{2}$ cup.

   $\frac{21}{32}$ cup mustard. Rounded off to $\frac{20}{30} = \frac{2}{3}$ cup.

   $\frac{15}{32}$ quart ketchup. Rounded off to $\frac{16}{32} = \frac{1}{2}$ quart.

   $1\frac{1}{8}$ quarts vinegar

   $\frac{57}{64}$ quart water. Rounded off to $\frac{56}{64} = \frac{7}{8}$ quart.

   $\frac{1}{8}$ cup pepper flakes
5. $\frac{1}{4} = \frac{15}{60}$ I got. He got $\frac{4}{15} = \frac{16}{60}$. He got 1/60 more.
6. 177 ounces for a dozen footballs; $262\frac{1}{2}$ ounces for a dozen basketballs. A dozen basketballs weigh $85\frac{1}{2}$ ounces more than a dozen footballs.
7. $\frac{1}{11}$ from Canada
8. $1\frac{1}{4}$ pounds of small red potatoes

   $\frac{5}{32}$ cup of ranch dressing. Rounded off to $\frac{4}{32} = \frac{1}{8}$ cup.

   $\frac{5}{8}$ teaspoon salt. Rounded off to $\frac{4}{8} = \frac{1}{2}$ teaspoon salt.

   $\frac{5}{64}$ teaspoon garlic powder. Rounded off to $\frac{4}{64} = \frac{1}{16}$ teaspoon.

   $3\frac{3}{4}$ slices of bacon. Rounded off to 4 slices.

   $6\frac{1}{4}$ black olives. Rounded off to 6 black olives.

   $\frac{5}{12}$ cup chopped onion. Rounded off to $\frac{1}{2}$ cup.
9. 9 days; $\frac{7}{10}$ mile to walk on the 9th day.

10. No. 10 children could get $4\frac{1}{2}$ inches of ribbon; 3 inches left over.
11. 36 grams of protein
12. $40\frac{8}{15}$ servings
13. I cannot make it to the mall in 3 hours. $\frac{13}{20}$ mile more to go.
14. 37 servings that are each exactly $5\frac{1}{3}$ ounces; $5\frac{1}{6}$ ounces of pie left over.

ANSWERS FOR Decimals

Review Exercise

1. 4 hundred thousands
   0 ten thousands
   6 thousands
   1 hundreds
   9 tens
   2 ones

2. 40,325,810

3. 762,400

Unit 1 Place Value, chapters The Place Value System and Place Values for Larger Numbers

4. 8,000

5. 7,600

6. 28,000

Unit 1 Place Value, chapter Rounding Numbers

7. Answers will vary, e.g. $\frac{10}{16}$, $\frac{20}{32}$

Unit 4 Fractions: Adding & Subtracting, chapter Equivalent Fractions and Lowest Terms

8. $1\frac{3}{7}$

Unit 5 Fractions: Multiplying & Dividing, chapter More About Dividing by Fractions

9. $\frac{9}{32}$

10. 105

Unit 2 Division, chapters Zeros in the Quotient and Three-Digit Divisors

Exercise 1

1. $\frac{4}{10}$,
   four tenths or point four

2. $\frac{8}{10}$,
   eight tenths or point eight

3. $1\frac{7}{10}$
   one and seven tenths or one point seven

4. 25, twenty-five

5. $\frac{5}{100}$, five hundredths or point zero five

6. $\frac{72}{100}$, seventy-two hundredths or point seven two

7. $11\frac{59}{100}$, eleven and fifty-nine hundredths or eleven point five nine

8. $17\frac{7}{100}$, seventeen and seven hundredths or seventeen point zero seven

9. 0.8

10. 8.1

11. 0.05

12. 0.77

13. 1.77

14. 36.01

15. 300.03

16. 10.10

17. .5

18. 1.1

19. 20.2

20. 0.09

21. 0.10

22. 0.65

23. 48.21

24. 500.04

25. 370.8

26. 999.39

27. =

28. >

29. <

30. >

31. <

32. >

33. >

34. <

35. <

36. <

Additional Exercises

1. $\frac{2}{10}$, two tenths or point two

2. $\frac{3}{10}$, three tenths or point three

3. $\frac{5}{10}$, five tenths or point five

4. $1\frac{6}{10}$, one and six tenths or one point six

5. $5\frac{9}{10}$, five and nine tenths or five point nine

6. 12, twelve

7. $\frac{7}{100}$, seven hundredths or point zero seven

8. $\frac{12}{100}$, twelve hundredths or point one two

9. $\frac{49}{100}$, forty-nine hundredths or point four nine

10. $6\frac{27}{100}$, six and twenty-seven hundredths or six point two seven

11. $9\frac{98}{100}$, nine and ninety-eight hundredths or nine point nine eight

12. $15\frac{4}{100}$, fifteen and four hundredths or fifteen point zero four

13. 0.5
14. 0.1
15. 25.9
16. 17.6
17. 0.07
18. 0.33
19. 0.81
20. 1.47
21. 1.01
22. 324.16
23. 10.10
24. 10.1
25. 204.04
26. 0.3
27. 0.9
28. 8.2
29. 23.3
30. 1,500.6
31. 0.05
32. 0.86
33. 154.14

34. 200.02

35. 770.07

36. 6.05

37. 707.07

38. 800.08

39. 100

40. 0.01

41. 2,536.5

42. =

43. >

44. <

45. =

46. >

47. >

48. >

49. >

50. <

51. <

Exercise 2

1. $\frac{3}{1,000}$, three thousandths or point zero zero three

2. $\frac{666}{1,000}$, six hundred sixty-six thousandths or point six six six

3. $7\frac{502}{1,000}$, seven and five hundred two thousandths or seven point five zero two

4. $1,000\frac{1}{1,000}$, one thousand and one thousandth or one thousand point zero zero one

5. $1,100\frac{1}{100}$, one thousand one hundred and one hundredth or one thousand one hundred point zero one

6. $\frac{82}{10,000}$, eighty-two ten-thousandths or point zero zero eight two

7. $100\frac{2,958}{10,000}$, one hundred and two thousand nine hundred fifty-eight ten-thousandths or one hundred point two nine five eight

8. $6\frac{561}{10,000}$, six and five hundred sixty-one ten thousandths or six point zero five six one

9. 0.001

10. 4.325

11. 20.022

12. 0.0001

13. 5.5111

14. 100.0100

15. 0.76

16. 1,275.0004

17. 0.08755

18. 36.049

19. 61.1422

20. 0.0487

21. 487.010

22. 500.041

23. 0.541

24. 0.932

25. 900.032

26. 0.0183

27. 180.0003

28. 10,000

29. 10,000.00 can also be just 10,000

30. 100,000

31. .00001

32. 0.50

33. 0.75

34. 1.200

35. 3.5

36. 29.1

37. 36.0

38. 0.047

39. 7.500

40. 0.0400

Additional Exercises

1. $\frac{1}{1,000}$, one thousandth or point zero zero one

2. $\frac{746}{1,000}$, seven hundred forty-six thousandths or point seven four six

3. $9\frac{109}{1,000}$, nine and one hundred nine thousandths or nine point one zero nine

4. $12\frac{36}{1,000}$, twelve and thirty-six thousandths or twelve point zero three six

5. $2\frac{100}{1,000}$, two and one hundred thousandths or two point one zero zero

6. $2\frac{10}{100}$, two and ten hundredths or two point one zero

7. $2\frac{1}{10}$, two and one tenth or two point one

8. $\frac{1}{10,000}$, one ten-thousandth or point zero zero zero one

9. $\frac{8,674}{10,000}$, eight thousand six hundred seventy-four ten-thousandths or point eight six seven four

10. $25\frac{465}{10,000}$, twenty-five and four hundred sixty-five ten-thousandths or twenty-five point zero four six five

11. $300\frac{3,498}{10,000}$, three hundred and three thousand four hundred ninety-eight ten-thousandths or three hundred point three four nine eight

12. $11\frac{7,645}{100,000}$, eleven and seven thousand six hundred forty-five one hundred-thousandths or eleven point zero seven six four five

13. 0.002

14. 0.349

15. 0.076

16. 5.003

17. 17.041

18. 0.0003

19. 0.5281

20. 0.0437

21. 1,000.0100

22. 0.052

23. 673.0007

24. 0.07671

25. 2.573

26. 345.0600

27. 0.169

28. 100.069

29. 700.037

30. 0.013

31. 13,000

32. 0.6421

33. 0.0005

34. 5.1350

35. 0.0857

36. 800.0057

37. 100.0003

38. 0.0103

39. 31

40. 31.0 could also just be 31

41. 31.00 could also just be 31

42. 2,531,600.76

43. 452,000

44. 400.052

45. 0.40

46. .9

47. 0.900

48. .0430

49. 1.00600

50. 0.8

51. .80000

52. 100.0

53. 1,000

54. 1.5

55. 1.50000

Exercise 3

1. 7.70
2. 8.16
3. 11.69
4. 6.756
5. 4.17
6. 3.049
7. 6.07
8. 8.301
9. 0.16
10. 1.87
11. 2.1964
12. 2.66
13. 5.3793
14. 0.951
15. 0.223
16. 0.07
17. 3.377
18. $14.23

19. $5.81

20. $48.80

21. 27.482 gallons left

Additional Exercises

1. 0.9
2. 1.5
3. 0.67
4. 8.45
5. 13.98
6. 2.59
7. 31.221
8. 4.57
9. 2.292
10. 8.47
11. 8.8194
12. 14.35
13. 5.103
14. 73.26
15. 0.08
16. 0.168
17. 1.86
18. 2.3395
19. 2.16
20. 2.202
21. 0.8
22. 3.4008
23. 8.315
24. 2.442
25. 0.48
26. 4.421
27. 3.376
28. 6.0185
29. $62.65
30. $35.25
31. $13.63
32. 5.626 gallons

Exercise 4

1. 0.45
2. 1.62
3. 48.00
4. 0.3267
5. 0.056
6. 0.00082

7. 27.29163

8. 0.042672

9. 1,117.805

10. 0.0091425

11. 361.595 miles

12. $17.10

13. $51.75

14. 347.066 pounds

Additional Exercises

1. 1.28
2. 3.54
3. 2.992
4. 5.136
5. 39.198
6. 0.0452
7. 227.9
8. 0.076
9. 447.9488
10. 0.00506
11. 0.000742
12. 12.42
13. 1.242
14. 0.00066
15. $34.68
16. $70.52
17. 37.725 gallons
18. 130.5105 pounds
19. 27,141.675 pounds

Exercise 5

1. 42.7
2. 7.5
3. 370
4. 648
5. 14,700
6. 590
7. 6,378
8. 0.5
9. 92,000
10. 825,932.7
11. 4,830,000
12. 10,000,000
13. $17.90
14. $99.00

15. 360 ounces
16. 5.6
17. 89.1
18. 0.705
19. 0.00823
20. 8.293
21. 0.000283

22. 38
23. 789.32
24. 15.32
25. 0.000056
26. 2.10 millimeters
27. 0.0025 meters
28. $21.10 per hour

Additional Exercises

1. 300
2. 32.6
3. 6.5
4. 1,034
5. 55,400
6. 1,710
7. 7,752
8. 0.1
9. 600
10. 60,200
11. 513,476.21
12. 7,140,000
13. 100,000
14. 10,000,000
15. $389

16. 3,090 pounds
17. 960 eggs
18. 3.9
19. 243.6
20. 0.103
21. 0.00741
22. 5.52
23. 17.496
24. 59
25. 1,245
26. 35.0007
27. 0.000104
28. 0.00082 meters
29. 8.67 pages per hour
30. $65 per hour

Exercise Dividing a Decimal by a Whole Number

1. .56
2. .007
3. 7.08
4. 10.8

Exercise Dividing a Whole Number

1. 3.5
2. 2.375
3. 2.875

Exercise Dividing a Decimal by a Decimal

1. .77
2. 3.4
3. 5.2

Exercise 6

1. 2.3
2. 2.25
3. 6
4. 6
5. 300
6. 180
7. 1.34
8. 34.5
9. 0.062
10. 0.0084
11. $12.50 per hour
12. 9.2 pounds

Additional Exercises

1. 1.12
2. 0.732
3. 4.3
4. 0.04

5. 1.2
6. 12
7. 120
8. 0.035
9. 0.0008
10. 30
11. 250.5

12. 5,260
13. 1.6875
14. 77.5
15. 17.5
16. 36 boxes
17. 73 hours

Exercise 7

1. 4.4
2. 4.39
3. 4
4. 62.7
5. 44.80
6. 45
7. 30
8. 110.48
9. 19,416.19

10. 8.39
11. 1.92
12. 4.20
13. 0.34
14. 22.4637 rounded to $22.46
15. 25.839793 rounded to 26 miles per gallon
16. 3.60833… rounded to 3.6 pounds

Additional Exercises

1. 7.2
2. 327.57
3. 8
4. 0.55
5. 1,359.1

6. 15
7. 20
8. 0.699
9. 100.0
10. 81.14

11. 1,974.73

12. 3.64

13. 4.74

14. 0.11

15. 0.16

16. 60.52

17. 5.1413881 rounded to 5.14 gallons

18. 44.44… boxes rounded to 44 boxes

19. $11.538017 rounded to $11.54 per hour

Exercise 8

1. 0.4
2. 0.75
3. 0.167
4. 0.438
5. 0.571
6. 2.25
7. 2.125
8. 1.25
9. 2.333
10. 0.263

Additional Exercises

1. 0.5
2. 0.8
3. 0.3
4. 0.111
5. 0.333
6. 2.2
7. 0.313
8. 3.333
9. 0.53
10. 0.583
11. 1.333
12. 20.833
13. 2
14. 0.214
15. 0.879

Exercise 9

1. $0.\overline{7}$
2. $0.\overline{18}$
3. $3.1\overline{6}$
4. $0.791\overline{6}$
5. $0.\overline{148}$

6. 0.27
7. 1.22
8. 2.06

Additional

1. $0.8\overline{3}$
2. $0.2\overline{7}$
3. $0.08\overline{3}$
4. $2.\overline{6}$
5. $0.\overline{259}$

6. $0.541\overline{6}$
7. 0.55
8. 2.35
9. 0.58
10. 6.92

Final Unit Exercises

1. 0.2   0.3   0.31
2. .627   6.2*   6.20*   6.27
   (*these are equal)
3. 0.0004   0.004   0.04*   0.040*   0.4
4. 2.0451   2.45*   2.450*   2.451
5. .0799   0.7*   .70*   1.00
6. 61.297   61.927   62.197
7. .15   .156   .16
8. .036   .063   .360
9. $\frac{621}{10,000}$
   six hundred twenty-one ten-thousandths or point zero six two one

10. $1{,}000\frac{108}{1{,}000}$,

    one thousand and one hundred eight thousandths or one thousand point one zero eight

11. $20\frac{0906}{10{,}000}$,

    twenty and nine hundred six ten-thousandths or twenty point zero nine zero six

12. $380\frac{420}{1{,}000}$,

    three hundred eighty and four hundred twenty thousandths or three hundred eighty point four two zero

13. 50.842 gallons

14. $197.78 rounded to the nearest hundredth

15. $209.33 rounded to the nearest hundredth

16. 14.324

17. .027

18. 3,006.3068

19. .000322

20. 15.2 ounces, rounded to the nearest tenth

21. 7.769 pounds left

22. 706.75 ounces

23. $10.\overline{15}$

24. 1.875

25. $.8\overline{3}$

ANSWERS FOR Metric Measurement

Review Exercise

1. a) $5\frac{9}{10}$, five and nine tenths or five point nine

   b) $\frac{3}{1,000}$, three thousandths or point zero zero three

   c) $100\frac{61}{100}$, one hundred and sixty-one hundredths or one hundred point six one

   d) $7\frac{502}{1,000}$, seven and five hundred two thousandths or seven point five zero two

   e) $1100\frac{1}{100}$, one thousand one hundred and one hundredth or one thousand one hundred point zero one

   f) $\frac{82}{10,000}$, eighty-two ten-thousandths or point zero zero eight two

2. a) 0.001
   b) 22.022
   c) 0.76
   d) 1,275.0004
   e) 36.049
   f) 61.1422
   g) 0.0487
   h) 487.010
   i) 500.041
   j) 0.541
   k) 100.7

   Unit 7 Decimals, chapters Writing Fractions Using the Place Value System and Thousandths and Ten-Thousandths

3. 9.0102

   Unit 7 Decimals, chapter Adding and Subtracting Decimals

4. 0.059486

   Unit 7 Decimals, chapter Multiplying Decimals

5. a) 10,150
    b) 217,000
    c) 0.1598
    d) 82,000
    e) 5,000

6. a) 0.1308
    b) 87.5
    c) 0.000417
    d) 0.00675
    e) 23.49

Unit 7 Decimals, chapter *Multiplying or Dividing Decimals by 10, 100, 1,000 and So On*

Exercise 1

1. 10 cm, 97 mm
2. 3 cm, 27 mm
3. 14 cm, 138 mm
4. 8 cm, 77 mm
5. 9 cm, 90 mm

Additional Exercises

1. 2 cm, 22 mm
2. 5 cm, 45 mm
3. 13 cm, 127 mm
4. 4 cm, 40 mm
5. 15 cm, 152 mm

Exercise 2

1. 2.7 cm, 27 mm
2. 7.8 cm, 78 mm
3. 14.3 cm, 143 mm
4. 5.4 cm, 54 mm
5. 1 cm, 10 mm
6. 225 mm
7. 986 mm
8. 101 mm
9. 327 mm
10. 180 mm
11. 98 mm
12. 8.8 cm
13. 0.3 cm
14. 3.0 cm

15. 3.5 cm

16. 1.9 cm

17. 4.5 cm

18.—28. Answers will vary.

29. 600 cm

30. 6,000 mm

31. .006 km

32. 720 mm

33. .72 m

34. .00072 km

35. .008 m

36. .000008 km

37. 800 cm

38. 147 cm

39. 71 m

40. .109 km

41. 2.03 m

42. 397 cm

43. 410 cm

44. .553 km

45. .1 cm

46. 12.2 cm

47. 1.52 m

48. 50 mm

49. 3.04 m

50. 351 km

51. 1.93 m

52. 1 mm

53. 90 m

Additional Exercises

1. 10.1 cm, 101 mm
2. 5.6 cm, 56 mm
3. 1.9 cm, 19 mm
4. 12.7 cm, 127 mm
5. 1.2 cm, 12 mm
6. 95 mm
7. 104 mm
8. 299 mm
9. 576 mm
10. 1164 mm
11. 3 mm
12. 6.6 cm
13. .8 cm
14. 2.0 cm
15. 4.5 cm
16. 1.3 cm

17. 8.5 cm

18.—29. Answers will vary.

30. 1,600 m

31. 160,000 cm

32. 1,600,000 mm

33. 500 cm

34. 5 m

35. .005 km

36. 4 m

37. 400 cm

38. 4,000 mm

39. 16.8 cm

40. .168 m

41. .000168 km

42. 250 cm

43. 500 mm

44. 764 mm

45. .764 m

46. .000764 km

47. 92 m

48. 37,400,000 mm

49. 172 cm

50. 7 mm

51. 135 cm

52. 1.55 m

53. .037 m

54. 220 cm

55. .00765 km

56. 3.3 m

57. 10,000 cm

58. .1 km

59. 80 cm

60. 1 cm

61. 18.3 m

62. 2.08 m

63. 216 mm

64. 3.05 m

65. 25.6 m

66. 1.63 m

67. 1,022 km

Exercise 4

1. 4,000 gm
2. 4,000,000 mg
3. 60.8 g
4. 60,800 mg
5. 2.149 kg
6. 2,149,000 mg
7. .0007 kg
8. 700 mg
9. .902 g
10. .000902 kg
11. 14,600 g
12. 1.002 kg
13. .006541 g
14. 50.4 mg
15. .551 g
16. 63 g
17. 709,000 mg
18. 12.576 kg
19. 18,000 mg
20. .020 g
21. 450 g
22. 50 kg
23. 5.67 g
24. 102 kg
25. 2,050 g
26. 3 g

Additional Exercises

1. 1,000 g
2. 1,000,000 mg
3. 500 g
4. 500,000 mg
5. 5,040 g
6. 5,040,000 mg
7. 1 kg
8. 1,000,000 mg
9. 2.039 kg
10. 2,039,000 mg
11. .0146 kg
12. 14,600 mg
13. .001 g
14. .000001 kg
15. .027 g
16. .000027 kg

17. 620 g

18. 4,100 g

19. .807 g

20. .991 kg

21. .0005 kg

22. 2.045 kg

23. 7,000 mg

24. 12.043 kg

25. 777,000 mg

26. 1 g

27. 5 g

28. 10,300 mg

29. .062 kg

30. 4,200 g

31. 6,060 mg

32. .097 g

33. 180 g

34. 2.27 kg

35. 1,272 kg

36. 1,400 g

37. 6 g

38. 500 g

39. 7 kg

40. 4 kg

41. 32 kg

42. 90 kg

Exercise 6

1. 1,300 ml

2. 70 ml

3. 601 ml

4. 22,030 ml

5. 1 L

6. .0351 L

7. .029 L

8. .4067 L

9. 47,000 ml

10. 5,090 ml

11. .006 L

12. .07009 L

13. 235 ml

14. 19 L

15. 40,000 L

16. 60 ml

17. 590 ml

18. 355 ml

19. 2,000 L

Additional Exercises

1. 1,000 ml
2. 1,250 ml
3. 10,000 ml
4. 10,060 ml
5. 40 ml
6. 1 L
7. .75 L
8. 5.167 L
9. .355 L
10. .014 L
11. 66,000 ml
12. 102,040 ml
13. .0946 L
14. 1,063 ml
15. .009 L
16. 22.496 L
17. 1,001,000 ml
18. 1.001 L
19. 87,000 ml
20. .087 L
21. 1.5 L
22. 1.89 L
23. 473 ml
24. 2 ml
25. 355 ml
26. 40 L

Final Unit Exercises

1. 4.096 m = .004096 km
2. 10.7 cm
3. 34.8 cm
4. 630 g
5. .108 g = .000108 kg
6. 8,000 ml
7. 1.005 L

ANSWERS FOR
Customary Measurement

Review Exercise

1. 30
2. 5,260
3. 1.75

Unit 7 Decimals, chapter Dividing Decimals

4. 8
5. 1,359.1

Unit 7 Decimals, chapter Rounding Decimals

6. 1974.73

Unit 7 Decimals, chapter Multiplying Decimals and chapter Rounding Decimals

7. 60.52
8. 44 boxes
9. $11.54 per hour

Unit 7 Decimals, chapter Dividing Decimals and chapter Rounding Decimals

10. .53
11. .583
12. 1.333
13. 20.833
14. 0.214

Unit 7 Decimals, chapter Rounding Decimals and chapter Repeating Decimals

Exercise 1

12. 4"
13. 6"
14. 2"
15. 3"

16. $3\frac{1}{4}"$

17. $3\frac{1}{8}"$

18. $3\frac{3}{16}"$

19. $3\frac{1}{2}"$

20. $3\frac{3}{4}"$

21. $3\frac{6}{8}" = 3\frac{3}{4}"$

22. $3\frac{11}{16}$

Additional Exercises

12. 2"

13. 2"

14. 5"

15. 1"

16. $1\frac{1}{4}"$

17. $1\frac{1}{8}"$

18. $1\frac{2}{16} = 1\frac{1}{8}"$

19. 3"

20. $3\frac{1}{4}"$

21. $3\frac{2}{8} = 3\frac{1}{4}"$

22. $3\frac{3}{16}"$

Exercise 2

1. 5 feet; 60 inches
2. 99 inches; 8 feet 3 inches
3. 10 feet; 120 inches
4. $43\frac{1}{2}$ inches; 3 feet $7\frac{1}{2}$ inches
5. 9 feet; 3 yards
6. 3 feet 5 inches; 1 yard 5 inches
7. 5 feet 5 inches
8. 6 inches
9. 1 yard
10. 2 inches
11. 3 yards
12. 84 feet
13. 3 miles
14. 42 inches
15. 4 miles
16. 90 inches
17. 132 inches
18. 13,728 ft.
19. 23 ft.

20. 15 yards
21. 8,976 yards
22. 26,928 ft.
23. 3 feet 10 inches
24. 1 foot $7\frac{7}{8}$ inches
25. 127 inches

Additional Exercises

1. 3 feet; 36 inches
2. 72 inches; 6 feet
3. 5 feet; 60 inches
4. 46 inches; 3 feet 10 inches
5. $97\frac{1}{2}$ inches; 8 feet $1\frac{1}{2}$ inches
6. 12 feet; 144 inches
7. $131\frac{1}{2}$ inches; 10 feet $11\frac{1}{2}$ inches
8. 2 feet
9. 6 inches
10. 1 yard
11. 2 yards
12. 10 feet
13. 100 yards
14. 16 feet
15. 10,560 feet
16. 7 yards
17. 47 yards
18. 3,168 yards
19. 9,504 feet
20. 31 ft.
21. 4.3 mi.
22. 1,800 inches
23. 18 feet 5 inches
24. 2 ft. $10\frac{1}{16}$ inches
25. 191 inches
26. 13 feet
27. $326\frac{3}{8}$ inches
28. 1 yard 30 inches
29. 33 feet 5 inches
30. 8 yards 14 inches
31. 33 yards 2 feet

26. $156\frac{5}{8}$ inches
27. 2 yards 23 inches
28. 5,333 yards 1 foot
29. 4 yards 29 inches
30. 42 feet 3 inches

Exercise 4

1. 64 ounces
2. 8.5 tons
3. 9 lbs.
4. 8,600 lbs.
5. 39 oz.
6. 4 lbs. 15 oz.
7. 7.4 pounds
8. 9 lbs. 6 oz.
9. 2.62 T
10. 160.96 oz.
11. $117\frac{1}{2}$ oz.
12. 1 lbs. $10\frac{3}{4}$ oz.
13. 2 lbs. $9\frac{1}{4}$ oz.
14. 1 lb.
15. 7 lbs.
16. 260 lbs.
17. 8 oz.
18. 25 lbs.

Additional Exercises

1. 128 oz.
2. 5 lbs.
3. 12,000 lbs.
4. 2.5 or $2\frac{1}{2}$ T
5. 30 oz.
6. $51\frac{1}{2}$ oz.
7. 3 lbs. 2 oz.
8. 8.5625 lbs.
9. 8 lbs. 9 oz.
10. 5.785 T
11. 32,180 lbs.
12. 11 lbs. 0 oz.
13. $219\frac{3}{4}$ oz.
14. .6875 pounds
15. .3125 pounds
16. 7 oz.
17. 5 lbs.
18. 15 lbs.
19. 60 lbs.
20. 200 lbs.
21. 3 lbs.

Exercise 5

6. 4 cups in a quart, 32 fluid ounces in a quart

7. 8 pints in a gallon, 4 pints in a half gallon

8. 64 fluid ounces in a half gallon

Additional Exercises

10. 2 cups in a pint, 16 fluid ounces in a pint

11. 4 cups in a quart, 32 fluid ounces in a quart

12. 2 pints in a quart

13. 4 quarts in a gallon

14. 4 pints in a half gallon, 8 pints in a gallon

Exercise 6

1. $4\frac{1}{2}$ cups
2. 44 fluid ounces
3. 16 pints
4. 8 cups
5. 40 qts.
6. $2\frac{1}{2}$ gal.
7. 448 fluid ounces
8. 20 cups
9. $\frac{3}{4}$ gal.
10. $\frac{1}{2}$ qts.
11. 30 gallons
12. 20 fluid ounces
13. 12 fluid ounces
14. $\frac{1}{2}$ fluid ounce
15. 3 gallons

Additional Exercises

1. 32 fluid ounces
2. 2 pints
3. 2 cups
4. 1 pint
5. 8 quarts
6. 2 gallons

7. 4 pints

8. $\frac{1}{2}$ gallon

9. 64 fluid ounces

10. 32 cups

11. $12\frac{1}{2}$ gallons

12. $1\frac{1}{2}$ gallons

13. 1 cup

14. $\frac{1}{2}$ gallon

15. 2 gallons

Final Unit Exercises

1. 15 feet 1 inch

2. 1.35 miles

3. $161\frac{1}{2}$ inches

4. 123 ounces

5. 19.61 tons

6. 76 cups

7. 4.5 quarts

ANSWERS FOR Positive and Negative Numbers

Review Exercise

1. a) $\frac{4}{21}$

    b) $\frac{1}{3}$

    Unit 4 Fractions: Adding & Subtracting, chapter Equivalent Fractions and Lowest Terms

2. $13\frac{27}{28}$

    Unit 5 Fractions: Multiplying & Dividing, chapter Multiplying Mixed Numbers

3. $3\frac{39}{64}$

    Unit 5 Fractions: Multiplying & Dividing, chapter More About Dividing by Fractions

4. 3.94

    Unit 7 Decimals, chapters Changing Decimals and Fractions, and Rounding Decimals

5. .063 kilometers

6. 24,060 millimeters

7. 7,100 meters

8. 0.164 meters

    Unit 8 Metric Measurement, chapter Changing From One Unit to Another

Exercise 1

1. integer
2. integer
3. not an integer
4. not an integer
5. integer
6. integer
7. integer
8. not an integer

9. not an integer
10. integer
11. integer
12. integer
13. integer

Additional Exercises

1. integer
2. integer
3. integer
4. integer
5. integer
6. not an integer
7. not an integer
8. integer
9. not an integer
10. not an integer
11. integer
12. not an integer
13. not an integer
14. not an integer
15. integer
16. integer
17. not an integer
18. not an integer

Exercise 2

1. 6
2. −1
3. 9
4. −1
5. 5
6. −6
7. 8
8. 6
9. 4
10. 0

Additional Exercises

1. 2
2. 10
3. 3
4. 1
5. 1
6. 2
7. −4
8. −1
9. 0
10. −5
11. 6
12. 0
13. 10
14. −4

Exercise 3

1. 7
2. 7
3. 12
4. 7
5. 4
6. 1
7. 2
8. 5
9. 0
10. 1
11. −1
12. −2
13. 0
14. 1
15. 2
16. −1
17. −3
18. 10
19. 0
20. 1
21. −1
22. −3
23. −6
24. 4
25. 6
26. −4
27. −3
28. 3

29. –6
30. –1
31. –2
32. 6

33. –3
34. –2
35. 1
36. –6

Exercise 4

1. –2
2. –5
3. 1
4. 1
5. –9

6. 6
7. –6
8. –6
9. –4
10. –10

Additional Exercises

1. 1
2. 3
3. 0
4. –1
5. –5
6. –6
7. –8
8. 0
9. –4
10. 9

11. –8
12. 0
13. –1
14. –7
15. –2
16. –6
17. 0
18. –1
19. –9
20. –10

Exercise 5

1. 7
2. 6
3. 5
4. −1
5. −3
6. −4
7. 2
8. 0
9. −1
10. −5
11. −2
12. −1
13. −5
14. −7
15. −4
16. −8
17. −9
18. 5
19. −9
20. −1
21. −7
22. −9
23. −4
24. 0
25. −9
26. −6
27. −3
28. 5
29. −7
30. −3
31. −2
32. 3
33. −10
34. −10
35. −1
36. −1

Exercise 6

1. −2
2. −5
3. 1
4. −6
5. −9
6. 6

7. −6
8. −9
9. −4
10. −10

Additional Exercises

1. 7
2. 5
3. 0
4. −1
5. −4
6. −10
7. −8
8. 0
9. 3
10. −2
11. −5
12. −7
13. −5
14. −3
15. 0
16. −10
17. −4
18. −4
19. −8
20. −6

Exercise 7

1. 2
2. 4
3. −3
4. −6
5. 3
6. −7
7. −7
8. −6
9. −2
10. −3
11. −9
12. −9
13. 1
14. 1
15. −7
16. −7
17. 11
18. −1

19. −1
20. −11
21. −11
22. 14
23. −2
24. −2
25. 2
26. −14
27. −14
28. 7
29. −7
30. −5
31. −8
32. −5
33. −2
34. 1
35. 1
36. −10

Additional Exercises

1. 3
2. −1
3. −3
4. 4
5. 2
6. −2
7. 1
8. 0
9. −1
10. −4
11. 2
12. −9
13. −5
14. 2
15. −2
16. −8
17. 2
18. 4
19. 1
20. 0
21. −1
22. 10
23. 1
24. 1
25. 2
26. 2
27. 0
28. 0

29. 1
30. 1
31. −1
32. −9
33. −8
34. −8
35. −5
36. −5
37. −7
38. −7
39. −1
40. −1
41. 1
42. −5
43. −10
44. 0
45. 5
46. −1
47. −8
48. −3
49. −9
50. −4
51. 2
52. −10

Exercise 8

1. 6
2. 4
3. 4
4. 10
5. −4
6. 8
7. 0
8. 2
9. −4
10. −1
11. 0
12. 1

Additional Exercises

1. 5
2. 5
3. 9
4. 2
5. 0
6. 1

7. 3
8. −1
9. 0
10. 1
11. 9

12. −1
13. 2
14. 9
15. −3
16. 9

Exercise 10

1. −2
2. −3
3. −5
4. 7
5. 6
6. 8
7. −10
8. 10
9. −2
10. 4
11. −9
12. 5
13. −5

14. −7
15. 2
16. −3
17. −7
18. −5
19. 8
20. −2
21. −9
22. 0
23. 9
24. 1
25. −7
26. −10

Additional Exercises

1. 18
2. 0
3. 0

4. 18
5. 0
6. −18

7. −18
8. 0
9. 5
10. 1
11. 5
12. 1
13. −5
14. −5
15. −1
16. −1
17. −6
18. −6
19. 2
20. 2
21. 6
22. −2
23. −2
24. 6
25. 8
26. −8
27. −8
28. 8
29. 5
30. −2
31. 4
32. −10
33. −16
34. 13
35. 6
36. 1

Exercise 11

1. −10
2. −9
3. −4
4. −6
5. 6
6. 6
7. 16
8. −16
9. −16
10. 0
11. 21
12. −24
13. −40
14. 0

15. −9
16. 1
17. −20
18. −20
19. 20
20. −20
21. −18
22. 30

23. 44
24. −42
25. 0
26. −35
27. −49
28. 49
29. −72
30. −18

Additional Exercises

1. 6
2. −6
3. −6
4. 6
5. 8
6. −8
7. −8
8. 8
9. 15
10. −15
11. −15
12. 15
13. 6
14. −6
15. −6

16. 6
17. 0
18. 0
19. 0
20. 0
21. 10
22. −18
23. 28
24. 36
25. −25
26. 0
27. −28
28. −10
29. 10
30. −24

31. 0
32. −40
33. 42
34. −27
35. 30
36. −14
37. 10
38. −40
39. −10
40. −42
41. −28
42. −4

Exercise 12

1. 2
2. 9
3. −21
4. 60
5. −4
6. −20
7. 14
8. 10
9. −7
10. −3
11. −32
12. 36
13. 0
14. −6
15. −45
16. 0
17. 5
18. −1
19. 36
20. −9
21. −48
22. −15
23. −8
24. 40
25. −11
26. 12
27. −4
28. 13
29. −70
30. −5
31. −22
32. 6

Additional Exercises

1. 6
2. 4
3. 4
4. 6
5. 5
6. −5
7. −4
8. −6
9. −6
10. −4
11. −5
12. 5
13. 7
14. 1
15. 1
16. 7
17. 12
18. −12
19. −1
20. −7
21. −7
22. −1
23. −12
24. 12
25. 2
26. −3
27. 6
28. 10
29. −9
30. 18
31. −16
32. −9
33. 6
34. −24
35. −3
36. −2
37. 3
38. −9
39. 11
40. −14
41. −5
42. −7
43. −20
44. 33
45. −1
46. −10

47. 0
48. 18
49. −32
50. 14
51. −40
52. −5
53. 6
54. 3
55. 40
56. 50
57. −56
58. −54
59. 0
60. −8

Exercise 13

1. 10
2. −10
3. −10
4. 10
5. 7
6. −5
7. −7
8. −8
9. 3
10. 4
11. −4
12. −3
13. 5
14. −8
15. 4
16. −2
17. 2
18. −3
19. −6
20. 5
21. 0
22. −7
23. −17
24. 6
25. 7
26. −7

Additional Exercises

1. 3
2. −3
3. −3
4. 3
5. 5
6. −5
7. −5
8. 5
9. −5
10. −5
11. 5
12. 5
13. −4
14. −4
15. 4
16. 4
17. 8
18. 0
19. 6
20. 7
21. −7
22. 5
23. −5
24. −5
25. −7
26. 2
27. 4
28. −7
29. −6
30. −5
31. −6
32. 4
33. −4
34. −9
35. 6
36. −7
37. 0
38. −4
39. −4
40. −1

Final Unit Exercises

1. 25
2. 2
3. 2
4. 3
5. −8
6. 2
7. −8
8. −7
9. −8
10. 16
11. 7
12. −6
13. −14
14. −7
15. 10
16. 2
17. −56
18. −5
19. −3
20. −2
21. 18
22. 3
23. −40
24. 5
25. 7
26. −4
27. 6
28. −2
29. −2
30. −2
31. −7
32. −63
33. −9
34. −5
35. 2
36. −8
37. −2
38. −20
39. 60
40. 8
41. 9
42. −32
43. 10
44. −10

45. 40
46. −64
47. −8
48. 3
49. 11
50. −7
51. 0
52. −9

Additional Exercises

1. 2
2. 5
3. −6
4. −2
5. −1
6. 10
7. 9
8. 6
9. −2
10. 13
11. −18
12. 2
13. 15
14. −3
15. −9
16. −12
17. −5
18. 35
19. −10
20. −4
21. 11
22. 4
23. −14
24. −4
25. −1
26. −42
27. −3
28. 45
29. 6
30. −5
31. −24
32. −7
33. −18
34. 0
35. −72
36. −6
37. 6
38. −6

39. 7

40. 0

41. –8

42. –5

43. –9

44. –2

45. –10

46. –1

47. –49

48. –3

49. 44

50. –5

51. –48

52. –12

ANSWERS FOR Simple Algebra

Review Exercise

1. 0
   Unit 10 Positive and Negative Numbers, chapter Subtracting Negative Integers

2. −5
   Unit 10 Positive and Negative Numbers, chapter Dividing Positive and Negative Integers

3. −742
   Unit 10 Positive and Negative Numbers, chapter Adding Negative Integers

4. −1,162
   Unit 10 Positive and Negative Numbers, chapter Multiplying Positive and Negative Integers

5. 24 pizzas, $408

6. 965 customers

7. 97.33
   Unit 3 Word Problems: Whole Numbers, chapter Some Math Definitions

8. $\frac{7}{12}$

9. $\frac{3}{5}$

10. $\frac{2}{3}$
    Unit 4 Fractions: Adding & Subtracting, chapter Equivalent Fractions and Lowest Terms

Exercise 1

1. $x + 4$
2. $p - 8$
3. $m + 3$
4. $9 + w$
5. $x - 6$
6. $6 - x$
7. $\frac{h}{5}$
8. $\$5,000 - x$
9. $\frac{\$10,000}{p}$
10. $d + 6$

Additional Exercises

1. $r + 9$
2. $x - 10$
3. $14 - m$
4. $m - 14$
5. $\$250 - d$
6. $r + 22$
7. $\frac{30}{n}$
8. $x - \$50$
9. $\frac{k}{10}$
10. $7.1 + x$

Exercise 2

1. 22
2. 7
3. 31
4. 49
5. 54
6. 39
7. 19
8. 15.3
9. 214
10. 0

Additional Exercises

1. 21
2. 32
3. 21
4. 16
5. 43
6. 1.9
7. 77
8. 6
9. 323
10. 3.204
11. 0

Exercise 3

1. $2x
2. $18
3. 5x
4. 55 minutes, 100 minutes
5. $\frac{1}{2}s$
6. $25
7. $5 + p$
8. $80, $42.65
9. $y - $10
10. $80
11. $25 + $5x
12. $55, $30
13. $\frac{m}{4}$
14. $25, $37.50
15. $3.50x
16. $10.50, $35.00
17. $8.50 + $4x
18. $16.50, $28.50
19. $\frac{$15}{x}$
20. 3 boxes, 5 boxes

Additional Exercises

1. $y + 7$
2. 11 years old
3. $50 - n$
4. 35 degrees, 49 degrees
5. $3p$
6. 42 points, 3 points
7. $m + 0.18
8. $1.93, $2.17
9. $1.50 + $3.50y$
10. $12.88, $20.75
11. $30 - x$
12. $22.25, $26.01
13. $\frac{n}{6}$
14. 17 cookies, 13 cookies
15. $8.50 + r$
16. $16.49
17. $250 - s$
18. $222.61
19. $\frac{$1,500}{w}$
20. 6 months
21. $40 + 16.5x$
22. $163.75

ANSWERS FOR SIMPLE ALGEBRA

Exercise 4

1. 7
2. 8
3. 16
4. 21
5. 5
6. 1,521
7. 0
8. 0.3125
9. 0.125
10. 0.148

Additional Exercises

1. 9
2. 19
3. 2.5
4. 32
5. 36
6. 0.67
7. 44
8. 0.5
9. 6
10. 2.19
11. 0

Exercise 5

1. −5
2. −4
3. 1
4. −8
5. 44
6. −18
7. −24
8. 7.5
9. 12
10. −14.6

Additional Exercises

1. −1
2. 13
3. −7
4. −45

5. 21

6. 11

7. −23

8. −0.17 (rounded)

9. 0

10. −12.15

Exercise 6

1. $.43 - .20 = b$
   $.23 = b$ salt left

2. $97s = 3,880$
   $s = 40$ seats in each row

3. $625 = \frac{p}{8}$
   $p = \$5,000$ lottery prize

4. $\frac{115.2}{x} = 4$ or $4x = 115.2$
   $x = 28.8$ kilograms for each

5. $0.39 = \frac{k}{5}$
   $k = 1.95$ kilograms total bought

Additional Exercises

1. $x = \frac{3.5}{2}$
   $x = 1.75$ tons for each

2. $467 = 78 + e$
   $e = 389$ not drivers

3. $3p = 66$
   $p = 22$ years

4. $39.24 = 3.6d$
   $d = 10.9$ kilograms

5. $\frac{f}{7} = 5.08$
   $f = 35.56$ kilograms

6. $c + 3c = 64$ or $4c = 64$
   $c = 16$ cans Ray collected

Final Unit Exercises

1. $17n$

2. $\$435 - n$

3. $\frac{x}{14}$

4. $n + 52$

5. $9 - 3n$ or $3n - 9$

6. $\$38.50(h) - \$25$

7. $\frac{q}{4}$

8. $\$1.75 + \$4.25y$

9. $\frac{w}{12}$

10. $290q$

11. $\$1.75(2) + \$4.25(4) = \$20.50$

12. $\$7,916.67$ (rounded)

13. $x = -0.081$ (rounded)

14. $y = -2.114$

15. $r = -25.326$

16. $m = -4.61$

17. $m - \$57.99 = \$209.56$
    $m = \$267.55$

18. $\dfrac{b}{2,000} = 3$
    $b = 6,000$ bottles

19. $20,000e = \$4,800$
    $e = \$0.24$ for each egg

ANSWERS FOR Ratio, Proportion and Percent

Review Exercise

1. Answers will vary.
   Examples: $\frac{1}{3}$, $\frac{6}{18}$

   Unit 4 Fractions: Adding & Subtracting, chapter Equivalent Fractions and Lowest Terms

2. $2\frac{11}{12}$

   Unit 4 Fractions: Adding & Subtracting, chapter Adding and Subtracting Mixed Numbers

3. $2\frac{5}{9}$

   Unit 5 Fractions: Multiplying & Dividing, chapter Multiplying Mixed Numbers

4. $4\frac{2}{3}$

   Unit 5 Fractions: Multiplying & Dividing, chapter More About Dividing Fractions

5. 10,070

6. 51,900

7. 0.405

8. 25.58

9. 0.00504

10. 0.023905

    Unit 7 Decimals, chapter Multiplying or Dividing Decimals by 10, 100, 1,000 and So On

11. .80

12. 3.875

13. .933 (rounded)

    Unit 7 Decimals, chapter Changing Decimals and Fractions

14. $\frac{9}{20}$

15. $\frac{4}{5}$

16. $\frac{83}{100}$

    Unit 7 Decimals, chapter Writing Fractions Using the Place Value System

Exercise 1

1. 2 circles to 6 squares
2. 3 oranges to 1 lemons
3. 7 blueberries to 5 cherries
4. 6 paddles to 4 people

Additional Exercises

1. 4 tennis balls to 2 basketballs
2. 1 table to 4 chairs
3. 10 books to 3 globes
4. 6 books to 4 tennis shoes

Exercise 2

1. 28 bowls
2. 40 chairs
3. 3 oranges
4. 9 turtle bowls
5. 3 triangles to 4 squares
6. 10 circles to 9 triangles
7. 6 squares to 5 circles
8. 5 circles to 6 squares
9. 12 eggs

Additional Exercises

1. $\frac{3}{4} = \frac{6}{8} = \frac{9}{12} = \frac{12}{16} = \frac{15}{20} = \frac{18}{24}$
2. $\frac{8}{1} = \frac{16}{2} = \frac{24}{3} = \frac{32}{4} = \frac{64}{8} = \frac{80}{10}$
3. 15 bowls
4. 35 chairs
5. 7 oranges
6. 12 turtle bowls
7. 54 students
8. 7 triangles to 12 squares
9. 12 squares to 7 triangles
10. 4 squares to 5 circles
11. 5 circles to 4 squares
12. 15 circles to 7 triangles
13. $\frac{18 \text{ customers}}{2 \text{ servers}} = \frac{9 \text{ customers}}{1 \text{ server}} = \frac{27 \text{ customers}}{3 \text{ servers}}$
14. $\frac{7 \text{ cups}}{4 \text{ days}} = \frac{14 \text{ cups}}{8 \text{ days}} = \frac{21 \text{ cups}}{12 \text{ days}} = \frac{28 \text{ cups}}{16 \text{ days}} = \frac{35 \text{ cups}}{20 \text{ days}} = \frac{42 \text{ cups}}{24 \text{ days}}$

Exercise 3

1. 4 people
2. 6 pounds
3. 1 chair
4. 10.5 miles
5. 9.6 minutes
6. 3.44 pounds (rounded)
7. 128 shots
8. 54 boys
9. 8 liters
10. 11.4 weeks (rounded)
11. 417 cars (rounded)
12. 64.5 kilometers (rounded)
13. 172 miles (rounded)

Additional Exercises

1. 9 cups
2. 1 pound
3. 37 tables
4. 6.5 hours
5. 210 minutes
6. 132 pages
7. 5.5 hours
8. 71.4 minutes (rounded)
9. $1,050
10. 32 pounds
11. 10 hours
12. 56.8 miles
13. $2.37 (rounded)

Exercise 4

1. 240 hours
2. 18 cups
3. $.24
4. $6\frac{3}{4}$ hours
5. 19 shots (rounded)
6. a) wingspan .55 meters
   b) length .5 meters
7. 8.15 hours (rounded)

Additional Exercises

1. 245 points
2. 136 people
3. .24 pounds
4. a) 6 buckets of cement
   b) 12 buckets of sand
   c) 18 buckets of gravel
5. 7 hours
6. 86.44 km./hr.
7. 2.5 feet (rounded)

Exercise 5

1. 60%
2. 80%
3. 4%
4. 100%
5. Answer given
6. 10%
7. $\frac{3}{10}$, 30%
8. $\frac{3}{5}$, $\frac{60}{100}$
9. $\frac{40}{100}$, 40%
10. $\frac{4}{5}$, $\frac{80}{100}$
11. $\frac{1}{4}$, 25%
12. $\frac{75}{100}$, 75%
13. $\frac{3}{20}$, $\frac{15}{100}$
14. $\frac{7}{20}$, 35%
15. $\frac{5}{100}$, 5%
16. 1, $\frac{100}{100}$
17. $\frac{2}{25}$, 8%
18. $\frac{45}{100}$, 45%
19. 45%
20. 83%
21. 37%
22. a) 70% red
    b) 30% not red
23. 40%
24. 20%
25. 98.9%
26. 85%
27. $31\frac{1}{4}$%

Additional Exercises

1. $\frac{3}{10}$, 30%
2. $\frac{9}{100}$, 9%
3. $\frac{11}{20}$, 55%
4. 25%
5. 40%
6. 5%
7. 10%
8. 1%
9. Answer given
10. $\frac{10}{100}$, 10%
11. $\frac{30}{100}$, 30%
12. $\frac{20}{100}$
13. $\frac{2}{5}$, $\frac{40}{100}$
14. $\frac{3}{5}$, 60%
15. $\frac{9}{20}$, 45%
16. $\frac{75}{100}$, 75%
17. $\frac{5}{100}$
18. $\frac{3}{20}$, $\frac{15}{100}$
19. $\frac{1}{4}$, 25%
20. $\frac{7}{20}$, 35%
21. $\frac{55}{100}$, 55%
22. $\frac{4}{5}$, 80%
23. a) 51% girls
    b) 49% not girls
24. a) 39% boys
    b) 61% not boys
25. 50%
26. 25%
27. 75%
28. 60%
29. 25.9%
30. 18%
31. 40%
32. $87\frac{1}{2}$%
33. 62.2%
34. 83.9%, no

Exercise 6

1. 12
2. 16
3. 75
4. $97\frac{1}{2}$
5. 90
6. 90
7. 2.24
8. 4.96
9. 487.5
10. 18.5
11. 3.78
12. 13.65
13. 35.625
14. 112.19
15. 50 pounds
16. 152 questions
17. 518 students
18. 138.89
19. 291 (rounded)
20. 529 (rounded)
21. 7 (rounded)

Additional Exercises

1. 20
2. 13.8
3. 200
4. $262\frac{1}{2}$
5. 120
6. 162
7. 6.08
8. 4.26
9. 1,225
10. 20.25
11. 6.44
12. 12.6
13. 49.6
14. 128.45
15. 175 people
16. 1,473 people
17. 1,875 apples
18. 25%
19. $4,987.50
20. 7.2
21. 700
22. 192

Exercise 7

1. $141.35

2. $1,620

3. a) $1.07 sales tax
   b) 7.9%

4. a) $.04 sales tax
   b) 4.5%

5. a) $89.44 discount
   b) $469.56 sale price

6. a) $749.50 discount
   b) $6,745.50 sale price

7. a) 6.7% discount
   b) $7,000 sale price

8. a) $24.50 discount
   b) 16.3% discount
   c) $135.59 total to pay

Additional Exercises

1. a) $3.85 sales tax
   b) $58.85 total paid

2. $90.10 total paid

3. $6.53 total paid

4. $.86 total paid

5. a) $4.20
   b) 7% sales tax

6. a) $.33
   b) 10% sales tax

7. $49.00 sale price

8. a) $84.00 discount
   b) $616.00 sale price

9. a) $86.93 discount
   b) $828.07 sale price

10. a) 24% discount
    b) $72.20 sale price

11. a) $17.49 sale price
    b) 30% discount

12. a) $10.00 discount
    b) 12.5% discount

Exercise 8

1. a) $100 interest
   b) $2,100 to pay off the loan

2. a) $120 interest
   b) $6,120 to pay off the loan

3. a) $845.04 interest
   b) $8,088.24 to pay off the loan

4. a) $1,060.42 interest
   b) $11,060.42 to pay off the loan

5. a) $8,859.13 interest
   b) $37,208.35 to pay off the loan

6. a) $7,268.43 interest
   b) $24,729.43 to pay off the loan

7. $.54 interest

Additional Exercises

1. a) $140 interest
   b) $1,140 to pay off the loan

2. a) $576 interest
   b) $2,176.00 to pay off the loan

3. a) $72 interest
   b) $1,672 to pay off the loan

4. a) $12 interest
   b) $1,612 to pay off the loan

5. a) $330 interest
   b) $2,530 to pay off the loan

6. a) $358.75 interest
   b) $8,558.75 to pay off the loan

7. a) $4,646.96 interest
   b) $14,295.96 to pay off the loan

8. $39,700 to pay off the loan

Exercise 9

1. 3.7%
2. 7.7%
3. 440%
4. 42.5 minutes
5. 29%
6. 9.9%
7. 27 games

Additional Exercises

1. 50%
2. 26%
3. 12.2%
4. $660,800
5. 300%
6. 75%
7. 49,000
8. 8.7%
9. 12.6%

Final Unit Exercise

1. △ △ △ ◯ ◯

2. $0.17, $0.12, the store selling 30 for $3.59

3. 16.7%

4. 43.8%

5. 35

6. 45 questions

7. $56.53

8. $15,070.50

9. 427%

ANSWERS FOR Simple Geometry

Review Exercise

1. 12 hours 10 minutes

   Unit 12 Ratio, Proportion and Percent, chapter Proportion

2. $0.12/ounce

   Unit 12 Ratio, Proportion and Percent, chapter Common Uses of Ratios and Proportions

3. 14%

   Unit 12 Ratio, Proportion and Percent, chapter Percent

4. $2,239.65

   Unit 12 Ratio, Proportion and Percent, chapter Common Uses of Percent

5. $12 - 6x$

   Unit 11 Simple Algebra, chapter Variables and Expressions

6. $a + 20 = 32$, $a = 12$

   Unit 11 Simple Algebra, chapter Addition and Subtraction Equations

7. $\frac{2,000}{d}$

   Unit 11 Simple Algebra, chapter Variables and Expressions

8. -25

9. -45

10. -7

    Unit 11 Simple Algebra, chapter Solving Equations

Exercise 1

1. 65°
2. 90°
3. 90°
4. 115°
5. 115°
6. 54°

7. 126°

8. 180°

9. 180°

10. 180°

11.
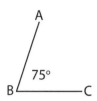

12.

Additional Exercises

1. 24°
2. 24°
3. 45°
4. 135°
5. 45°
6. 135°
7. 180°
8. 90°
9. 102°
10. 78°
11. 42°
12. 12°
13. 138°
14. 90°
15. 180°
16. 180°
17. 180°

18.

19.
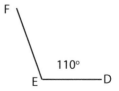

Exercise 2

1. 12 cm
2. 36.8 cm
3. 7 ft $11\frac{3}{4}$ in
4. 31 ft 4 in or 10 yd 1 ft 4 in
5. 21 cm
6. 5 in
7. 14 cm
8. 21 cm
9. 17 m, $187
10. 12 m

Additional Exercises

1. 19.2 cm
2. 22 cm
3. 14 ft $1\frac{3}{4}$ in
4. 22 ft 8 in
5. 17.4 cm
6. 6 in
7. 15.5 cm
8. 8 cm
9. 29 m, $290
10. 14 m

Exercise 3

1. (answers vary)
2. 37.68 m
3. 31.85 m
4. diameter 71.34 m, radius 35.67 m
5. 26.5 cm
6. 9 ft 2 in
7. 1.59 km
8. 3.34 cm

Additional Exercises

1. 15 cm
2. 14 in
3. 47.1 m
4. 21.18 m
5. 11.93 cm
6. 5.97 cm
7. diameter 23.89 cm, radius 11.95 cm
8. 9 ft 11 in
9. 6.53 km

Exercise 4

1. 2 dimensions
2. 2 dimensions
3. 3 dimensions
4. 1 dimension
5. 2 dimensions
6. 2 dimensions
7. 3 dimensions
8. 2 dimensions

Additional Exercises

1. 2 dimensions
2. 2 dimensions
3. 2 dimensions
4. 2 dimensions
5. 1 dimension
6. 3 dimensions
7. 3 dimensions
8. 3 dimensions
9. 3 dimensions
10. 1 dimension

Exercise 5

1. BC is parallel to ED, ED is perpendicular to LH, AK is perpendicular to FG
2. AH is parallel to BG, DJ is parallel to EI, CF is perpendicular to DJ and EI

Additional Exercises

1. JB is parallel to IC, AF is perpendicular to HD
2. FB is parallel to GA, CE is perpendicular to HD

Exercise 6

1. 64
2. 64
3. 3,125
4. 1
5. 100
6. 1,000
7. 6,561
8. 81
9. 729
10. 4

11. 27

12. 512

13. 64

14. 81

15. 625

16. 128

17. 9^5

18. 10^7

19. 4^3

20. 15^2

21. 50^3

22. 13^6

Additional Exercises

1. 36
2. 100
3. 16
4. 125
5. 144
6. 1,000
7. 1,024
8. 256
9. 49
10. 27
11. 216
12. 81
13. 144
14. 1,000
15. 81
16. 3,125
17. 512
18. 1
19. 7^4
20. 10^6
21. 15^3
22. 20^2
23. 14^3
24. 6^5
25. 9^4
26. 343

Exercise 7

1. area 9.1 cm², perimeter 16.6 cm
2. area 55 ft², perimeter 32 ft or 10 yd 2 ft
3. area 60 cm², perimeter 37 cm
4. 144 in²
5. area 2,340 in², perimeter 16 ft 10 in
6. 10,000 cm² (1 m²)

Additional Exercises

1. area 36.75 m², perimeter 24.8 m
2. area 36 ft², perimeter 30 ft or 10 yd
3. area 42 mi², perimeter 26 mi
4. area 25.2 cm², perimeter 85.2 cm
5. area 88.5 m², perimeter 51 m
6. 9 ft²
7. 6.45 cm²

Exercise 8

1. 3 m²
2. 18 ft²
3. area 24 cm²
4. area of ABCD 60 m², perimeter 38 m, area of ABC 30 m²
5. P = h + w + h + w
6. A = s x s or s²

Additional Exercises

1. 1,000 yards²
2. 17.5 m²
3. 3.25 cm²
4. area of ABCD 112 cm², perimeter 44 cm, area of BCD 56 cm²
5. area 54 m², perimeter 36 m
6. perimeter = s + s + s + s, or 4s
7. perimeter = c + b + h

Exercise 9

1. 44 ft^2
2. 14.88 m^2
3. 7 m^2
4. 11.34 cm^2
5. 1.13 m^2
6. 78.5 ft^2
7. area 8.04 km^2, circumference 10.05 km, diameter 3.2 km, radius 1.6 km
8. area 211.13 cm^2, circumference 51.5 cm, diameter 16.4 cm, radius 8.2 cm

Additional Exercises

1. 60 cm^2
2. 473 mm^2
3. 45 ft^2
4. 186.44 m^2
5. 153.86 m^2
6. 2.54 cm^2
7. 0.38 km^2
8. area 186.17 m^2, circumference 48.36 m, diameter 15.4 m, radius 7.7 m
9. area 1,610.89 m^2, circumference 142.24 m, diameter 45.3 m, radius 22.65 m
10. area 0.025 m^2, circumference 0.57 m, diameter 0.18 m, radius 0.09 m

Exercise 10

1. 348 ft^2
2. 2,628 mm^2
3. 79.38 m^2
4. 138.16 ft^2
5. 93.96 m^2
6. 345.4 cm^2

Additional Exercises

1. 62 in²
2. 344.44 cm²
3. 26.28 cm²
4. 835.24 cm²
5. 181.24 m²
6. 640.56 cm²

Exercise 11

1. 125.31 cm³
2. 4,176 mm³
3. 343.36 m³
4. 102 ft³
5. 75.36 m³
6. 156 cm³
7. 1,728 in³
8. volume 25 ft³, weight 1,560 pounds

Additional Exercises

1. 58.8 cm³
2. 240,000 ft³
3. 339.12 cm²
4. 150 m³
5. 80 in³
6. 1,356.48 m³
7. 1,000,000 cm³
8. 3,931.2 pounds

Exercise 12

1. 6
2. 8
3. 11
4. 14
5. 17
6. 5.57
7. 31.62
8. 100
9. 1.5
10. 0.4
11. 105.8 m
12. 1,320 ft

Additional Exercises

1. 4
2. 9
3. 13
4. 18
5. 20
6. 25
7. 6.40
8. 30
9. 26.46
10. 200
11. 2.5
12. 0.71
13. 4.69 m
14. 11.68 m
15. 2,379.7 ft

Exercise 13

1. 7.07 cm
2. 18.87 ft
3. 24 m
4. 61.93 ft
5. 94.34 ft
6. 30 in

Additional Exercises

1. 8.49 cm
2. 15.81 ft
3. 11.998 m
4. 7 m
5. 11.60 cm
6. 33.11 m
7. 7.42 ft

Final Unit Exercise

1. 21°
2. 111°
3. 90°
4. 10 ft 11 in
5. 47.1 m
6. 3^5, 243
7. 22^2, 484
8. 8^3, 512
9. 4,235 in²
10. area 594 mm², perimeter 114 mm
11. area 122.66 m², fence = 39.25 m, cost = $588.75
12. cylinder surface area 628 m². Box surface area 688 m². The cylinder has a smaller surface area.
13. 12 cm

ANSWERS FOR Supplemental Exercise Sheets

Table of Metric Units			
length	1 km	=	1,000 m
	1 m	=	100 cm
	1 cm	=	10 mm
weight	1 kg	=	1,000 gm
	1 gm	=	1,000 mg
volume	1 L	=	1,000 ml

Table of Customary Units			
length	12 in.	=	1 ft.
	3 ft.	=	1 yd.
	5,280 ft.	=	1 mi.
weight	16 oz.	=	1 lb.
	2,000 lb.	=	1 T.
volume	8 fl. oz.	=	1 c.
	2 c.	=	1 pt.
	2 pt.	=	1 qt.
	4 qt.	=	1 gal.

Exercise Sheet 1

1. a) hundreds
   b) 700
2. Four hundred nine thousand seven hundred sixty-five
3. 8,400
4. 5 R 18
5. 420 gallons
6. $\frac{5}{6}$
7. $3\frac{23}{24}$
8. $2\frac{1}{4}$
9. $13\frac{1}{3}$
10. 7.70
11. .45

12. .4

13. See chart above.

14. a) .006 kg
    b) 6,000 ml
    c) 2 cm
    d) 1.5 m
    e) 1,500 m

15. See chart above.

16. a) 1 ft. 1 in.
    b) 11 yd. 0 in.
    c) 1 mi. 240 yd. 0 ft.
    d) 1 lb.
    e) 2 T.
    f) 2 c. 2 oz.
    g) 4 pt.
    h) 1 qt. 1 pt.
    i) 2 gal. 2 qt.

17. 3:45 or 3 hours and 45 minutes

18. $228.00

Exercise Sheet 2

1. a) ones
   b) 3

2. nine hundred fifty thousand four hundred forty-four

3. 1,000

4. 7 R 22

5. 185 people

6. $\frac{23}{52}$

7. $2\frac{2}{9}$

8. $\frac{11}{48}$

9. $2\frac{5}{8}$ cups

10. 2.66

11. 1.12

12. .571

13. See chart.

14. a) 9,800 gm,
    b) 2,500 ml,
    c) 900 mm,
    d) 1,000 cm,
    e) 2,000 m

15. See chart.

16. a) 6 yd. 2 ft.
    b) 8 yd. 1 ft.
    c) 1 mi. 1,240 yd. 0 ft.
    d) 5 lb. 8 oz.
    e) 3 T.
    f) 3 c. 0 oz.
    g) 3 pt.
    h) 3 qt. 1 pt.
    i) 1 gal. 2 qt.

17. 193 minutes, 3:13 or 3 hours and 13 minutes

18. $7

Exercise Sheet 3

1. 145,890

    1: hundred thousands
    4: ten thousands
    5: thousands
    8: hundreds
    9: tens
    0: ones

2. One hundred forty-five thousand eight hundred ninety

3. 55,000

4. 307

5. $228

6. $\frac{7}{8}$

7. $5\frac{5}{32}$

8. $1\frac{7}{15}$

9. 18% (rounded)

10. .135,678

    1: tenths
    3: hundredths
    5: thousandths
    6: ten thousandths
    7: hundred thousandths
    8: millionths

11. One hundred thirty-five thousand six hundred seventy-eight millionths

12. $.\overline{7}$

13. See chart.

14. a) 2.4 kg
    b) .24 l
    c) 2.5 cm
    d) 2.05 m
    e) 1.002 km

15. See chart.

16. a) 146 in.
    b) 186 ft.
    c) 31,697 ft.
    d) 19.2 oz.
    e) 13,000 lb.
    f) 148 fl. oz.
    g) 33 c.
    h) 49 pt.
    i) 22 qt.

17. 3:00 p.m.

18. $6.48

Exercise Sheet 4

1. 129,945,890

    1: hundred millions
    2: ten millions
    9: millions
    9: hundred thousands
    4: ten thousands
    5: thousands
    8: hundreds
    9: tens
    0: ones

2. One hundred twenty-nine million nine hundred forty-five thousand eight hundred ninety

3. 129,946,000

4. 706 R10

5. $62 more

6. $3\frac{8}{33}$

7. 11

8. $2\frac{14}{15}$

9. Furry Pets has 8 puppies, Pets Galore has 9 puppies

10. .412,502

    4: tenths
    1: hundredths
    2: thousandths
    5: ten-thousandths
    0: hundred-thousandths
    2: millionths

11. four hundred twelve five hundred two millionths

12. .1$\overline{48}$

13. See chart.

14. See chart.

15. a) 630 minutes
    b) 10 hr. 30 min.

16. a) 800 cupcakes
    b) 58 boxes

Exercise Sheet 5

1. 999,999,999

    9: hundred millions
    9: ten millions
    9: millions
    9: hundred thousands
    9: ten thousands
    9: thousands
    9: hundreds
    9: tens
    9: ones

2. Nine hundred ninety-nine million nine hundred ninety-nine thousand nine hundred ninety-nine

3. 1 billion or 1,000,000,000

4. 8

5. $11.25

6. $\frac{7}{40}$

7. $1\frac{1}{4}$

8. $\frac{5}{20} = \frac{1}{4}$

9. a) $\frac{1}{2}$
   b) 6 donuts

10. Forty-one thousand two hundred fifty hundred thousandths

11. 6

12. 59

13. See chart.

14. See chart.

15. 12 t-shirts

16. a) 3 bulbs for $4.29
    b) $12.30

Exercise Sheet 6

1. area = s x s or B x h
   (s = side, B = base, h = height)
   perimeter = side1 + side2 + side3 + side4
   Examples will vary.

2. area = $\frac{1}{2}$ x b x h (b = base, h = height)
   perimeter = side1 + side2 + side3
   Examples will vary.

3. volume = l x w x h (l = length, w = width, h = height) or B x h
   Examples will vary.

4. volume = Bh or $\frac{1}{2}$whl
   Examples will vary.

5. C = πd or C = π2r
   Examples will vary.

6. A = πr²
   Examples will vary.

7. 2x

8. x = 21

9. x = 7

10. x = −18

11. c − 12 = 25
    c = 37

12. $1,140 total paid back

13. $\frac{60}{3} = \frac{?}{12}$, 240 hours

14. $\frac{12 \text{ decrease}}{30 \text{ original}} = \frac{? \text{ \% decrease}}{100\%}$, 40%

15. 44

Exercise Sheet 7

1. area = wh
   perimeter = side1 + side2 + side3 + side4
   Examples will vary.

2. area = $\frac{1}{2}$ wh
   perimeter = side1 + side2 + side3
   Examples will vary.

3. volume = lwh or volume = Bh
   Examples will vary.

4. volume = Bh or volume = $\frac{1}{2}$ whl
   Examples will vary.

5. C = πd, or C = π2r
   Examples will vary.

6. A = πr²
   Examples will vary.

7. 2 (2 + 4 and 6)

8. 50 − n

9. t = 21

10. h = 21

11. m = −23

12. $\frac{3,880}{97}$ = s or 97s = 3,880
    s = 40 seats

13. a² + b² = c²,
    c = 7.07 cm

14. $\frac{6}{16} = \frac{?}{48}$, ? = 18

15. 50%

16. $141.35

17. 11%

18. -6

Exercise Sheet 8

1. 128 cm²
2. 96 cm³
3. 62 cm²
4. 30 cm³
5. 69.08 cm²
6. 15.7 cm³
7. x = 10 cm
8. 24 cm
9. −12.15
10. 2
11. 14 − m
12. m − 14
13. r + 22
14. P − 50
15. $\frac{p}{625} = 8$ or $\frac{p}{8} = 625$ or p = 8(625), $5,000
16. 54 boys
17. 12 cups
18. 20 miles (rounded to a whole number)
19. -72

Exercise Sheet 9

1. 8.5 cm (rounded)
2. 8.9 cm (rounded)
3. 114 cm²
4. 50 cm³
5. −45
6. 6.75 hr. ($6\frac{3}{4}$)
7. The surface area of the cylinder is 628 m². The surface area of the box is 688.6 m². The cost of the cylinder is less, so it is best.
8. .55 m
9. .5 m
10. $24,729.43
11. 2.5 kg, $\frac{x}{5} = .5$ or 5 x .5 = x
12. 2

Exercise Sheet 10—Vocabulary

1. square
2. rectangle
3. circle
4. parallelogram
5. right triangle
6. triangle
7. cube
8. cylinder
9. parentheses
10. fraction
11. square root
12. expression
13. equation
14. multiply
15. divide
16. subtract
17. add
18. equals
19. percent
20. divisor
21. dividend
22. quotient
23. numerator
24. denominator
25. Geometry is the part of mathematics that is about lines, angles and shapes.
26. Plane geometry is the part of geometry that studies things that are all in one plane.
27. Algebra is a part of math where letters or other symbols are used to stand for amounts in order to solve problems.
28. An integer is a positive or negative whole number (not a decimal or fraction). Zero is also an integer.
29. A ratio is a comparison of one amount to another.
30. A proportion says that two ratios are equal.
31. See chart.
32. See chart.

Printed in the USA
CPSIA information can be obtained
at www.ICGtesting.com
LVHW081548240524
781020LV00006B/38

9 780897 392006